Marshall Monroe Kirkman

The Compound Locomotive

Supplement to the Science of Railways

Marshall Monroe Kirkman

The Compound Locomotive
Supplement to the Science of Railways

ISBN/EAN: 9783744661942

Printed in Europe, USA, Canada, Australia, Japan

Cover: Foto ©Andreas Hilbeck / pixelio.de

More available books at **www.hansebooks.com**

THE COMPOUND LOCOMOTIVE

SUPPLEMENT TO

THE SCIENCE OF RAILWAYS.

BY

MARSHALL M. KIRKMAN.

THIS SUPPLEMENT IS RESPECTFULLY PRESENTED BY THE WORLD
RAILWAY PUBLISHING COMPANY TO SUBSCRIBERS TO
THE EIGHTH EDITION OF "THE SCIENCE OF RAIL-
WAYS" AND IS DESIGNED TO FORM A PART
OF THE "ENGINEERS' AND FIREMEN'S
MANUAL," VOLUME XII.

EIGHTH EDITION.

NEW YORK AND CHICAGO:
THE WORLD RAILWAY PUBLISHING COMPANY.
1899

CHAPTER I.

ENGINEERS' AND FIREMEN'S MANUAL, CONTINUED—
COMPOUND LOCOMOTIVES—INTRODUCTORY.

NOTE—In regard to the merits of the compound cylinder as compared with the single-expansion, I do not desire nor profess to express any opinion. I merely wish to describe the compound cylinder in what follows, and, if in some places preference seems to be expressed, it is the claim of builders and not mine. Those who use locomotives must themselves be the judges of the respective merits of single-expansion and compound cylinders and of the particular pattern they want.

January, 1899.   M. M. K.

In view of the fact that the Compound Locomotive is of comparatively recent introduction as compared with the single-expansion cylinder, its construction and working are less understood by those connected with the equipment department of railroads. These particulars I have, for that reason, thought it best to embody here in connection with the "ENGINEERS' AND FIREMEN'S MANUAL." More and more attention is being given to the expansive use of steam in connection with the locomotive. It is claimed not to be unreasonable in view of the triple and the quadruple expansion of steam (expansion in three and four cylinders) in the most approved stationary and marine engines that it should be possible to devise practical means of obtaining double expansion, at least, in a locomotive. Because of the more or less general introduction of compound cylinders, some account of the working of the compound locomotive is becoming every day more and more necessary to those connected with the machinery and equipment of railroads. Indeed, practical familiarity with the working of compound engines may be said to have become, in a measure, a necessary part of the knowledge of every engineer and fireman, for the reason that their duties may at any time, through promotion or otherwise, take them to roads where some form of compound locomotive is extensively used. Moreover, it is well that their attention be especially directed to the subject in order that it may have the consideration and scrutiny at their hands which its growing importance justifies and their practical knowledge is likely to render so valuable.

It is not surprising, inasmuch as the opinions of engineers and firemen respecting the operation of simple engines differ so widely, that there should be much controversy among them in regard to the operation of compound locomotives. Experience on the part of those operating the compound locomotive will tend to its better service and consequently greater development. Many prejudices against it are due to lack of acquaintance with its operation. This is true of every new thing. Especially is it true in the case of compounds where a railway has few locomotives of this kind. The feeling is but natural, if we remember, as we should, that in the handling of their engine the reputation of the engineer and fireman is at stake. The opportunity afforded them for handling and studying any odd engine, compound or otherwise, is necessarily limited. Moreover, in the operation of such locomotive, it is possible they may be impressed with the unhandy, because unusual, cab arrangements. Naturally, they are filled with apprehension, lest some accident might occur and they be found ignorant of what should be done to temporarily repair the engine and bring it in. If a considerable number of the locomotives of a railroad are compounds of similar class, the men handling them become accustomed to their operation and this fear disappears. It is with a view to the practical usefulness these pages may have for engineers, firemen, and others interested in the operation of compounds, that the descriptions herein are elaborated to the extent they are.

The plan followed with each class of compound locomotives is, first, to give a general description of its operation, succeeded by a detailed description of its technical parts. This is done that the reader may, in the first instance, if he desires, learn of the general arrangement of each class without the details, and afterwards, at his leisure, he may apply himself to the particular class or classes that most interest him. Following the description of each class of compounds will be found, in catechetical form, information relating to its operation in case of any ordinary derangement of its parts when the methods of procedure will differ from those in case of similar accidents to simple locomotives, as outlined in the manual. It is said that reforms must pass through three stages: ridicule, argument, and adoption. The compound has been subjected to the first and second of these epochs, and it may be assumed to have reached the last stage mentioned. Like every part of a railroad, it is still in a state of evolution. I know of no preference in regard to compounds that is proper to express here. The order of description has, therefore, no significance. In the description of the various types I have endeavored to eliminate all matter that does not pertain directly to the practical application of the principle of compounding, as descriptions relating to other parts of the locomotive are to be found elsewhere in "THE SCIENCE OF RAILWAYS." In relation to the descriptions of the compound, I wish to say that I am indebted in a marked manner to Mr. E. W. Pratt, whose familiarity with the construction and working of locomotives and the appliances of the latter makes him an authority of the highest order in regard to all such matters. I am indebted to him in many other ways and it affords me much gratification to be able to acknowledge it thus conspicuously.

CHAPTER II.

A compound locomotive is one in which the exhaust from one or more cylinders is passed into one or more other cylinders and made to do more work by further expansion before it is allowed to escape to the atmosphere. In stationary and marine service the principle of compounding has long since passed its experimental stage and, following the replacement of the single-expansion engine by the double-expansion type, came the era of high boiler pressures with triple and even quadruple-expansion engines in marine service. It was long thought by many, and is still held by some, that, although compounding of steam in marine and stationary engines was a great economy due to the use of the condenser,* on locomotives where condensers were impracticable, the compound locomotive would not be able to gain sufficient advantage over the simple engine to warrant their general use.

Without any attempt to pass judgment upon the relative value of the points put forward for and against compound locomotives, I will out-

---

*A condenser is a chamber into which the final exhaust of an engine takes place and in which the steam is cooled and condensed, either by a jet of water or by contact with sheets or tubes having cold water circulation on their opposite sides. These two forms of condensers are termed "jet condenser" and "surface condenser," respectively.

line some of the claims made by their advocates
and also some of the practical objections met
with in their use, many of which objections
have been largely overcome in the later
designs.

It should be remembered that the locomotive
is not a steam engine merely, but consists of a
boiler as well, and must also carry water and fuel
for its own demands.

The first advantage of the compound over the
simple locomotive comes from its greater economy
in fuel, resulting primarily from the saving in
steam. There is, however, a secondary saving, produced by the less violent effect of the exhaust upon
the fire and also the economical use of high boiler
pressures in compound engines.

Experiments have shown that high boiler pressures, say above 180 pounds, which have been
found very economical (especially in the space
occupied per horse power developed) in stationary and marine engines, are not a source of economy with the type of single-expansion locomotives
in use in this country, due to the extreme ranges
of temperature within a single cylinder and the
consequent condensation. Also, locomotive cylinders and their steam ports are not well protected,
and compounding the cylinders renders the variations of temperature in each cylinder less wide,
and thus makes practicable the use of higher pressure. With the use of steel in place of iron for
boiler construction, and also on account of the
excellent care and inspection given all locomotive boilers, there is no material increase of first

cost or for maintenance of high-pressure boilers. In simple locomotives the exhaust produces such a violent draft upon the fire that great quantities of unconsumed fuel are drawn from the fire-box and thrown from the stack. This is not alone a waste of fuel, but cinders entering the open car windows are a source of great annoyance to passengers, while, before the use of the compound locomotive in the service of suburban railways, the noise of the exhaust had to be overcome by means of mufflers, which became quickly choked up and produced a high back-pressure on the pistons, resulting in the loss of from 15 to 20 per cent. of the power.

The throwing of sparks from the stack is not only a source of annoyance, but frequently results in heavy losses from damage by fire in timber and agricultural districts, and this is largely, if not entirely, overcome by the compound locomotive.

The heating surfaces of any given boiler absorb heat from the fire and deliver it to the water at a certain rate. If the rate at which the products of combustion are carried away exceeds this rate of absorption, there will be a continual waste that can only be overcome by reducing the velocity of the products of combustion. In the compound locomotive this is effected by the milder exhaust. It has been clearly demonstrated by experiment, that a milder exhaust and consequently a slower rate of combustion produces a greater evaporation of water per pound of fuel.

The milder exhaust is, of course, the result of a

lower back-pressure and thereby permits a greater effective power on the piston.

There is also found to be considerable reduction in cylinder condensation, owing to the relatively small variation of temperatures in each cylinder as compared with single-expansion engines. In any engine the walls of the cylinder, one cylinder-head, and one side of the piston are cooled to the temperature of the exhaust steam during each stroke, and the live steam, again entering, must

Fig. 100.

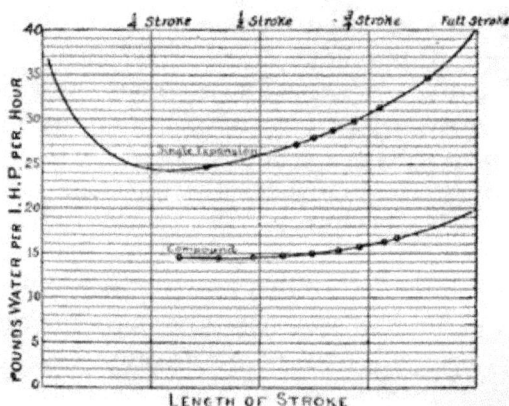

reheat them to its own temperature, thus condensing and requiring additional steam to flow in and take its place. In the compound this total range of temperature is divided between the high and the low-pressure cylinders, and thus the variation and consequent condensation in each cylinder is less.

The saving of steam results in the saving of

both water and fuel. The economy in fuel can be directly reduced to dollars and cents, while that resulting from the saving of water is more indirect. In bad water districts, the reduction of from 15 to 20 per cent. in the amount of water used, necessitates less frequent washings-out of the boiler and must result in greater life and diminished repairs to boiler and flues. Moreover, as its carrying capacity of water limits the distance that a locomotive can run without stopping (or slowing up, where track-tanks are used), it is evident that the compound locomotive would have an advantage in this respect. Fig. 100, besides showing the most economical point of cut-off for simple and compound engines, as far as the use of water is concerned, clearly shows the relatively smaller amount of water used by the compound per indicated horse power.

No locomotive can haul more than its adhesion to the rails will permit, and hence the tractive power of an engine is based upon whatever the adhesion to the rails may be. This is determined by practical experiment. With a fairly dry rail, a turning force of more than one-fifth, or 20 per cent., of the weight of the drivers on the rails, will cause the wheels to slip; a perfectly dry rail will permit of a tractive power of about one-fourth, or 25 per cent., of the weight on the drivers; a well sanded dry rail will allow one-third, or $33\frac{1}{2}$ per cent., while a bad frosty rail will permit less than half this last amount. Where all the driving wheels are connected, it matters not, of course, whether this force is ap-

plied by one or many cylinders, but, if the power
is not uniformly distributed throughout the revolution and becomes sufficiently excessive at any
one point to cause the wheels to slip, a very
much less power will thereafter keep them slipping.   It is a well known fact that adhesion, and
consequently the tractive power of a locomotive,
is very much reduced after the wheels begin to
slip.*

It is claimed for the compound, that, as the
average cut-off is later in both cylinders than for
simple engines, the turning power is more uniform throughout the revolution, and hence heavier
trains can be hauled than with the single-expansion engine.   Then, too, while it would be un-
economical at other times to design a simple
engine with cylinders sufficiently large to develop
so high a tractive power as $33\frac{1}{3}$ per cent. at slow
speeds, this can be done with compound locomotives of the "convertible" type without loss
in economy under ordinary speeds of service,
when working compound.

A saving of oil has been one of the minor econ-
omies claimed to be incidental to the use of compound locomotives.   It is generally thought that
from six to ten drops of valve oil per minute are
required to be supplied with the steam in order
to properly lubricate a valve and cylinder.   This

---

*Every engineer knows this and puts his knowledge into prac-
tice when on a very slippery rail by opening the throttle very
slightly and leaving the valve in full gear, thus distributing the
pressure more uniformly throughout the stroke than would be
the case with a shorter cut-off.

oil is supplied to the high-pressure cylinder only, and hence, in the two-cylinder class of compounds, a saving has been effected in many cases.

Comparative tests of greater or less duration have been made by various railways, between compound and simple locomotives of otherwise similar construction, and the results obtained by the different experimenters are widely at variance. In general, it may be said that the reported saving in fuel with the compound is about ten per cent. in fast passenger and 20 per cent. in heavy freight service, although figures double the latter have frequently been given.*

Later designs of compound locomotives, arranged to be worked simple at the will of the engineer, will temporarily pull a heavier train than a simple engine of otherwise like design. When it is considered that the ruling grade on a division is the governing factor for the maximum rating of through trains over the whole division, it will be seen that a locomotive capable of enough greater power to haul an additional car or two up that grade, produces more economical service for the whole division.

Leaky valves and cylinder packing are less wasteful in a compound than in a simple engine. Steam leaking by the valve or packing of the high-pressure cylinder is still worked expansively

---

*Fast freight service will more nearly compare with express passenger service and the saving will be less than in heavy slow freight service, on account of the simple engine using steam more expansively in the former service than in the latter. This is more fully brought out elsewhere.

in the low-pressure cylinder.   Then, too, the difference of pressure between the two sides of the valves and pistons is less than for simple engines, and the wear should be consequently less.

On the other hand many serious practical objections have been raised.   The large cylinders greatly increased the weight of the reciprocating parts.*   This must be followed by heavier counter-balance† weights and their accompanying evils.‡ Also larger ports and consequently larger valves must be provided for the large cylinders. Inasmuch as considerable difficulty was formerly experienced in obtaining admission and exhaust ports of sufficient size for the cylinders of large high-speed single-expansion engines, it is not remarkable that there should have been considerable trouble experienced from this source in designing ports for the much larger cylinders of the compound locomotive.

In connection with these last two points, the weight of reciprocating parts and the port re-

-----

*By the reciprocating parts is meant those parts that have a forward and back (or reciprocating) motion. This includes the pistons and piston-rods, the cross-heads and a certain part of the main rods.

†The counter-balance is the balance weight placed in the wheel at a point opposite the crank pin. (See No. 240, Plate I.)

‡The counter-balance weights act at all points of the revolution, having an outward or centrifugal tendency from the center of the wheel that is great at high speeds, and is only counteracted when the engine is passing the front and back centers. At other points it produces an upward tendency upon the engine when moving up and a downward blow upon the rails when moving down.

quirements, let us see for one moment what the requirements for high speed are. Take an engine with a five foot driving wheel and a twenty-four inch stroke, traveling at the not unusual rate of a mile a minute. This requires 336 revolutions per minute, and means that the piston starts and stops 672 times, and with all its stopping and starting travels 1344 feet per minute. At mid-stroke the piston speed is about 35 feet per second or 24 miles per hour. To the average railway man high speeds are so common that I can think of no better way to show the full meaning of these figures than to compare them with those of a falling body acted upon by gravity. A falling body at the end of the first second is traveling at the rate of 32 feet per second, or about 22 miles per hour, and in gaining this speed the body has fallen some 16 feet, through which distance the continued action of gravity has produced this not inconsiderable acceleration, while in the case of the piston of the locomotive in question, the still greater speed of 24 miles per hour must not only be attained by the time the piston has reached the middle of its stroke, or in one foot, but also it must be exerting great additional propelling power on the crank pins. The steam in front of this rapidly moving piston must be exhausted freely or there would be no effective power to maintain the speed. Where the reciprocating parts weigh nearly a thousand pounds, it will be seen what enormous power is required to stop and start them without jar or shock to the loco-motive, and also the size of ports required to

freely exhaust the pressure ahead of and freely
supply steam behind the very large pistons of
compound locomotives. With any of the several
types of valve gear used in locomotive practice
to-day, it goes without saying that no compound
can be relatively so efficient, in comparison with
single-expansion engines, in express passenger as
in freight service, for the reasons hereinbefore
described. When many single-expansion locomotives with moderately large ports require a velocity of steam through their ports of over 1,000
feet per second, it can better be imagined than
told what the port requirements are for the large
low-pressure cylinder on a compound. It is generally considered that for express passenger service the low-pressure cylinders should not be so
large in proportion to the high-pressure cylinders
as for freight service.

Many of the earlier compounds in this country
suffered in comparative tests with simple engines
of otherwise similar design, by having cylinders
of too small a size to do the same work as the
simple engine. While they did their work with
economy, they would not haul the heavy trains
of the simple engines, and their supposed capacity
had to be reduced, to the annoyance of those en-
gaged in the operating of trains. Since then, with
the advent of larger cylinders and the "convertible" class of compounds, the conditions have
become altered.

With any locomotive, when steam is shut off,
as in running down grade, the pistons act as air
compressors, causing thumping, rough riding, and

cooling of the cylinders, as well as a strong draft in the stack at a time when no steam and little draft are required, and thus produce a waste of fuel. The large low-pressure cylinders of the compound have greatly magnified this evil, and several builders have overcome it by the use of automatic valves on the low-pressure cylinder, by which the two sides of the low-pressure piston are connected whenever the locomotive is drifting.

After closing the throttle, an engine working compound will make several strokes before all steam is finally exhausted. This delay in clearing the cylinders of steam, placed compound locomotives to considerable disadvantage in switching or like service; also, in such service, accustomed to gauge the speed by the exhaust of the engine, trainmen were often deceived by the less frequent exhaust of the two-cylinder compound. The employment of the separate high-pressure exhaust, whereby the engine can be run simple at the will of the engineer, it appears, has overcome these objections from an operating standpoint.

Many of the earlier forms of intercepting valves were of the poppet type and hammered badly in opening or closing. It will be noticed that these valves in the later designs are of the piston type and are almost invariably cushioned by dash-pots connected thereto, or some other equally effective means.

It is also claimed that the breakage and loosening of the large low-pressure cylinders have been considerably done away with by the use of proper reducing valves and a better attachment of cylinders to the frame, the use of double front rails for the latter being particularly noticeable in modern construction of large locomotives.

Thus it is that improvements in design and construction are continually taking place, and the upholders of the great principle of compounding will certainly witness their more extensive adaptation to all classes of service.

CHAPTER III.

There are many classes of compound locomotives in use. First, the *strictly plain compound*, where no live steam is admitted to the low-pressure cylinder, even in starting. The Webb three-cylinder compounds (with cylinders arranged as outlined in Fig. 106,) which are usually without connecting rods—the two high-pressure cylinders turning the rear pair of driving wheels by two outside cranks while the low-pressure cylinder turns the forward drivers by means of an inside crank—belong to this class and are used in considerable numbers on the London & North-Western of England. They are not powerful in starting, as the driving wheels acted upon by the high-pressure cylinders must turn, either by slipping or moving the train, before steam enters the low-pressure cylinder.

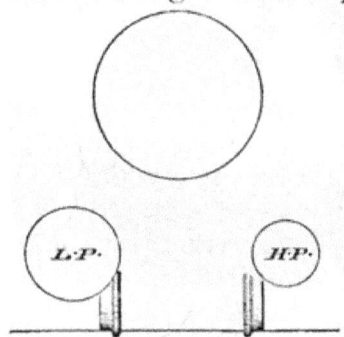

Fig. 101.

Second, *automatic compounds*—those using live steam in the low-pressure cylinder in starting only, automatically changing to compound with the first stroke, and thereafter cannot be run except as compounds.

(19)

The third class can be run simple or compound at any time at the will of the engineer and will be termed *convertible compounds.*

Each of these principal classes may have two, three, or four cylinders. The two-cylinder or "cross-compound" always has an intermediate receptacle, called a receiver,* between the high and low-pressure cylinders, while the four-cylinder engines may or may not have receivers — those with both pistons attached to the same crosshead generally have not. The three systems of four-cylinder compounds used in this country are the Baldwin (Vauclain), the Brooks (Player), and the John-

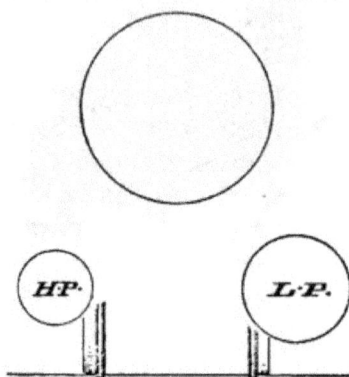

Fig. 102.

stone. Of these, the Brooks has receivers, while the remaining two are of the continuous expansion† type and have no receivers.‡

*The receiver is for the purpose of receiving the exhaust from the high-pressure cylinder and holding it until the engine gets to the point in the revolution where it is admitted against the low-pressure piston. Incidentally, the receiver may act as a re-heater, if located in the smoke-box, as is usually the practice.

†Meaning expansion without any pause or interruption as is the case when a receiver is interposed between the high and the low-pressure cylinders.

‡In Europe the Hungarian State Railways employ four-cylinder tandem compounds with one receiver into which both high-pressure cylinders exhaust and from which both low-pressure cylinders receive their supply.

The arrangement of cylinders is quite varied, as shown by the several skeleton cuts. There may be two cylinders, one high-pressure and one low-pressure, as outlined in Figs. 101, 102 and 103; one high and two low-pressure, as in Figs. 104 and 105; two high-pressure and one low into which they both exhaust, as in Fig. 106; two high, each exhausting independently into a low-pressure cylinder on the same side of the engine, shown in Figs. 107, 108, 109 and 110; or two high exhausting into a common receiver from which both low-pressure cylinders draw their supply, as in Figs. 111 and 112. Aside from the varied arrangement of cylinders, many of the European designs employ three and four cranks and use no side rods. Some French constructions, retaining the use of side rods, employ for the high-pressure cylinders two inside cranks on one driving axle at an angle with the low-pressure cranks on a second driving axle, the angle between the cranks

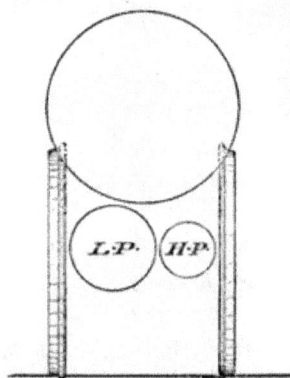

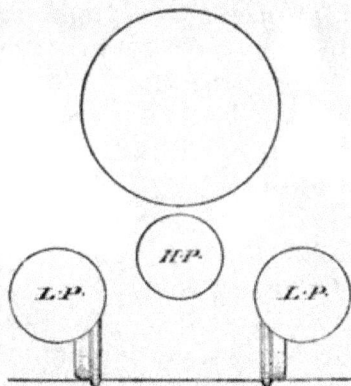

Fig.103.

Fig.104.

being such as to give as large a turning power as possible, for all portions of the revolution.

It is, perhaps, needless to say that the wide variations in the service of American locomotives demand that they have a large starting power at all points of the stroke. To obtain this starting power, all the earlier designs used a device called an intercepting valve that, if closed in starting, cut off communication between the receiver and the low-pressure cylinder and at the same time admitted

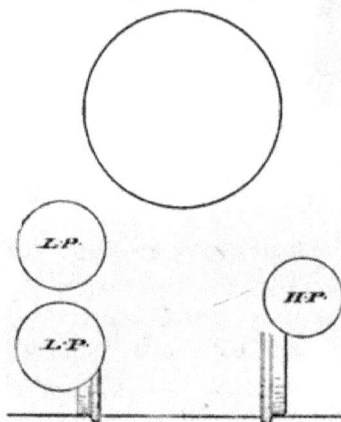

Fig. 105.

live steam to the low-pressure side, but after the first exhaust from the high-pressure cylinder to the receiver took place, the pressure in the latter automatically shoved open the intercepting valve and simultaneously s h u t off the further supply of live steam to the low-pressure cylinder. Hence these engines belong to the automatic class of compounds.

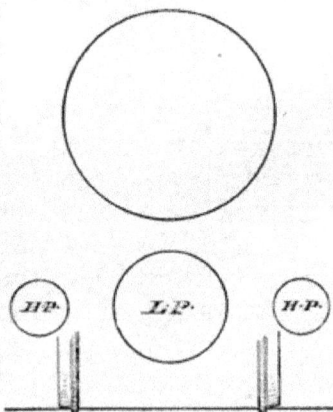

Fig. 106.

Mr. Anatole Mallet, who was the designer of the first practical compound locomotives in Europe

in 1876, was also the first to devise a means by
which a compound could be worked as a simple
locomotive for any desired period at the will of
the engineer. This was accomplished by adding
a separate exhaust valve through which the exhaust from the high-pressure cylinder could escape
to the atmosphere without accumulating in the
receiver. This relieved all back pressure on the
high-pressure piston and admitted of greater power
at slow speed than was otherwise obtained.*

Many objections were raised to placing the
operation of the engine
either as a simple or as a
compound in the hands
of the engineer, and the
fear was freely expressed
that the average engineer
would run the locomotive
to its disadvantage in
simple position more than
enough to offset the sav-
ing when operated as a
compound. However, one
prominent railroad officer,
in placing the operation of the valves at the will
of the engineer, seemed to express the now settled
conviction of all, when he said: "To argue that an

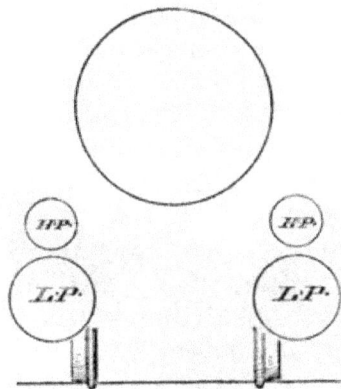

*It should also be stated that not only were the automatic
compounds less powerful, at slow speeds after starting, than
simple engines, but, except in the case of four-cylinder engines
having one high and one low-pressure cylinder on the same side
they were practically helpless in case of a broken steam chest
on either side. The use of the separate exhaust valve has greatly
altered the conditions in these cases.

engineer is likely to work simple any longer than
absolutely necessary, is about the same as saying
that an engineer with the ordinary engine cannot
be trusted to pull the reverse lever up as soon as
possible."

Later practice interposed, within or near the
intercepting valve, a reducing valve, which is
used to admit live steam, at a reduced pressure
only, into the low-pressure cylinder when starting
or when working simple. This reduced the
abnormal shocks that were
produced when starting
large compounds of earlier
design. The reducing
valve, the intercepting
valve, and the separate
exhaust valve were so
closely combined in many
cases and so dependent,
one upon the other, in
their operation, that it
became the tendency
among railway and me-
chanical men to refer to the whole mechanism
simply as the "intercepting valve."

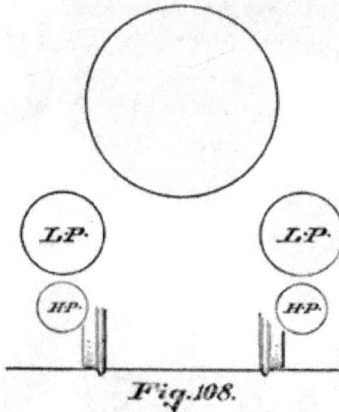

Fig. 108.

While the limit to the size of the ordinary
locomotive may be considered to have been
reached when the largest practical boiler that can
be placed on a given gauge track has been attained,
the limit to the American two-cylinder, or cross-
compound, with outside cranks will be the maximum width allowable for locomotives. However,
again Mr. Mallet, the father of the present era

of compound locomotives, has seemingly solved
the problem by dividing the low-pressure cylinder
into two cylinders, as shown in Fig. 105, of smaller
size attached to the same crosshead. With such
a construction it would appear that the boiler
would still be the limiting feature of the size of
the compound as well as the simple locomotive.

The proper cylinder ratio of compounds for all
varieties of service is still somewhat undeter-
mined. By the cylinder ratio is meant the proportion between the volumes of the high and the
low-pressure cylinders, not including the clearance

Fig. 109.

spaces. In American practice where the length
of stroke is the same, the cylinder ratio would
be as the areas of the two pistons, and it can
readily be found by multiplying the diameter of
each cylinder by itself and comparing the two
products. For instance, to find the cylinder ratio of an engine with a 20 inch high and a 30
inch low-pressure cylinder, multiply 20x20 equals
400; 30x30 equals 900; 400 goes in 900 two and
one-fourth times, which is the cylinder ratio.

The early practice in this country with two-cylinder compounds gave a ratio of two to one or even less, but extended experiment has dem-onstrated that a greater proportion than this is advisable a n d m a n y compounds of this class have a low-pressure cyl-inder from two and one-half to two and three-quarters times larger than their high-pressure cylinder. The Baldwin works used a ratio of 3 to 1 in their Vauclain four - c y l i n d e r com-

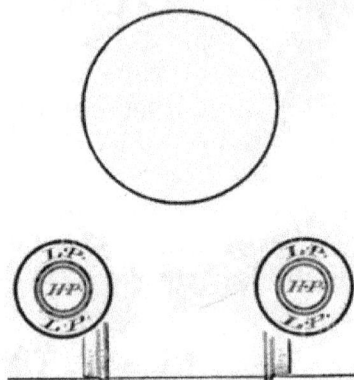

Fig. 110.

pounds for both passenger and freight service for a number of years and consider the results emi-nently satisfactory, while the builders of the Brooks tandem four - c y l i n d e r compound advise a ratio of between 2.8 and 3 to 1. However, the whole prob-lem of cylinder ratios for compound locomotives is based upon the desirabil-ity of dividing the work as equally as possible be-tween the high and the low-pressure cylinders, and

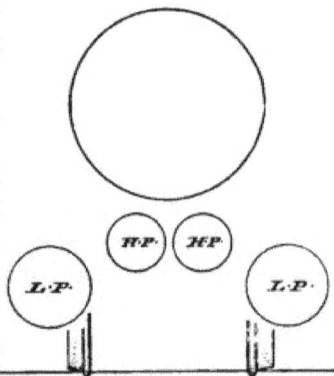

Fig. 111.

without going into details, it is apparent that no given ratio will keep the work equally divided for different service and different points of cut-off, nor should this equal division of power between the cylinders be given anything but secondary consideration in comparison with the total economy of the locomotive. To partially equalize the power of compounds, the amounts of lap and lead are not the same for both cylinders; one builder uses a separate lever in the cab for independently adjusting the travel of the low-pressure valve, as fully described elsewhere.

There seems to be no general rule followed by builders in this country as to which cylinder of a two-cylinder compound should be placed on the right-hand or engineer's side of the engine. Generally the intercepting valves are located on the engineer's side to make their connections as simple as possible, and hence, according as the design contemplates the placing of this valve adjacent to the high or the low-pressure cylinder, that one is placed on the right-hand side. But even this rule is not without exception.*

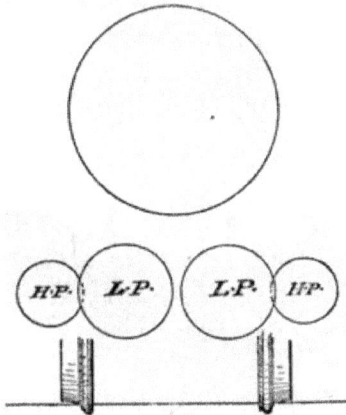

Fig.12.

----

*It would seem as though the intercepting valve, if placed between the high-pressure cylinder and the receiver, would cause less wire-drawing of steam to the low-pressure cylinder than if located between the latter and the receiver

Some compounds have cylinder casings both of the same size, but with the advent of the thirty-four or thirty-five inch low-pressure cylinder it seemed to many advisable to place it on the engineer's side with the thought of its better protection from damage if within his vision, and, furthermore, that the high-pressure cylinder casing be made no larger than necessary for reason of its better protection from accident.

It is becoming the general practice on compounds of any size to place combination safety and relief valves on the receiver and the low-pressure chest and cylinder heads to avoid damage in case of broken reducing valve or other accident that might produce unsafe pressure on that side.

CHAPTER IV.

In what respects should the instructions for firing compound locomotives differ from those for single-expansion engines, as described in the Manual? On account of the effect of the milder exhaust on the fire, the fireman should carry the fire as light as possible and yet keep a thin layer of coal in a well ignited state over the entire surface of the grates.

Should the engineer observe any difference between the operation of a compound and a simple locomotive? Perhaps the most important thing is that the engineer must accustom himself to use the reverse lever for what it is intended, not hesitating to increase the valve travel whenever necessary, and converting the engine to "simple" only at low speeds and after the maximum power of the engine working compound is found insufficient. There will be no such waste of steam from lack of expansion (see Fig. 100) nor such serious injury to the fire by working the valves full gear as is the result with single-expansion engines. Do not expect to find the best running position of the reverse lever as near the center of the quadrant as with a simple engine. The engine will do better work with the lever nearer full gear, just how far depending, of course, on the varying conditions of train, grade, and speed; but, in general, do not cut the lever back too far. The engine throttle should usually be run as wide open as possible, but judgment must be used in this respect according to the existing circumstances.   An engine designed for heavy freight service may do better under very light work or in passenger service, if throttled.

In order to move and stop engines quickly when desired, as for engine-house or turn-table work, how is it advisable that compound engines be handled?   Their valves should be kept in simple or starting position whenever their construction will permit, in order that the engines may be moved with the least necessary opening of the throttle and that, when the throttle is closed, the pressure remaining in the pipes will the sooner escape.

In starting, why are compounds more likely to slip than other engines? If they are started working simple, their power temporarily is considerably greater than that of simple engines of the same weight, hence great care should be exercised by the engineer to avoid slipping; but at any rate the cylinder cocks should be left open as long as there is any liability of water being in the cylinders.

Why is it, that, under these circumstances, compound locomotives frequently slip only a part of a turn instead of spinning around, as many simple engines will do unless steam is immediately shut off? Because, while they are more powerful at slow piston speeds, the ports used for starting are generally so small that a rapid movement of the piston will so greatly reduce the effective pressure on the piston as to cause the slipping to cease, many times without closing the throttle.

What harm results from frequent slipping for only a part of a revolution? This will occur each time at that part of the stroke where the turning power is the greatest and finally wear the tires out of round, so that the wheels will pound badly at high speeds, thus causing damage to the machinery and track besides making the engine ride hard.

Wherein does the lubrication of their cylinders differ from that of simple engines? Oil need only be supplied to the high-pressure cylinder or cylinders, when the engine is working compound. Hence a four-cylinder compound would be lubricated, as usual, on both sides of the engine, but a two-cylinder, or cross-compound, would require oil only in the one high-pressure cylinder when working in compound position. In this latter class, valve oil should be supplied to the low-pressure cylinder, however, when the engine is starting or running simple, but immediately shut off as soon as the change is made to compound working.

What are apt to be the injurious consequences of feeding oil to the low-pressure side when the engine is working compound? There being then no supply of live steam through the reducing and the intercepting valves, the oil would settle there and gum these valves so as to cause them to stick and, possibly, become inoperative.

When is it advisable to work the compound as a simple engine? In starting very heavy trains, and also on heavy grades when it is seen that the power as a compound with the valve in full gear is insufficient.

When the engine can be run simple for any period at the will of the engineer, above what speed is it impracticable to do so? Above a speed exceeding six or eight miles an hour.

What is the reason for this? Because, as outlined elsewhere, the special ports for use when the compound locomotive is working simple are usually purposely designed too small to permit of much speed.

If special valves for use when the engine is drifting are not provided, what is it advisable to do? Work a little steam all the time in order to keep the large valves and pistons somewhat lubricated.

Why is it more difficult to detect leaky valves and cylinder packing in compound than in simple engines? In the first place it does not so seriously affect the good working of the engine, and, in the second place, because the high-pressure cylinder does not exhaust to the atmosphere where a blow could be heard, and also the pressure behind the low-pressure piston is very much lower than with a simple engine and, consequently, a blow does not sound as loud.

What general rule will apply to compound as well as simple locomotives where one side has to be disconnected in cases of accident? On the disabled side, always block securely at both ends of the crosshead and firmly secure the valve, under all ordinary circumstances, in the center of its seat?

What is usually the cut-off in the high-pressure cylinder with the reverse lever hooked up? Some builders cut no notches higher up in the quadrant than would give a cut-off at one-half stroke, while others notch the quadrant all the way up to the center.

Why is it not necessary to have notches in the quadrant for shorter cut-off than one-half? A shorter cut-off than one-half for a compound locomotive is generally inadvisable for reasons analagous to those hereinbefore urged against a shorter cut-off than one-fourth for a simple locomotive; also it prevents damage that might ensue from excessive compression. With a ratio of 2 to 1 between the sizes of the low and the high-pressure cylinders, one-half cut-off gives practically four expansions, a three-fourths cut-off gives two and two-thirds expansions; with a cylinder ratio of 3 to 1, one-half cut-off gives about six, and three-fourths cut-off about four expansions. Fig. 100 clearly shows, that, while the most economical cut-off (as far as the amount of steam used is concerned) is about one-quarter of the stroke for single-expansion engines, it is about one-half stroke for compounds. Lighter work than is produced by full throttle and one-half cut-off in the high-pressure cylinder can best be obtained by throttling the engine.

In what way does the use of a separate exhaust valve give greater power to the two-cylinder compound at slow speeds? Assuming 200 pounds boiler pressure and a cylinder ratio of $2\frac{1}{2}$ to 1, the high-pressure exhaust reaches about 57 pounds when the engine is working compound. This pressure is the receiver pressure and becomes the initial pressure (at slow speeds) in the low-pressure cylinder. When working simple, the reducing valve permits about 80 pounds of live steam to pass into the low-pressure cylinder, thus considerably increasing the power in that cylinder. The high-pressure piston also exerts an equally increased power when working simple, as the back pressure of 57 pounds is removed by the separate exhaust valve.

THE BALDWIN FOUR-CYLINDER COMPOUND.

The builders of the "Vauclain" four-cylinder compound locomotives claim a design productive of the greatest efficiency with the utmost simplicity of parts and the least possible deviation from existing practice; that they also develop equal power on each side of the locomotive, thereby preventing the racking of the machinery resulting from an unequal distribution of power; and that, in their method of handling by the engineer, there is but slight departure from that of single-expansion or non-compound locomotives. They may be started, and run for any desired length of time, either simple or compound, at the will of the engineer, and can be changed from the one to the other at his discretion by the movement of a small lever in the cab which also operates the cylinder cocks.

The principal features of construction are as follows:

The cylinders consist of one high-pressure and one low-pressure cylinder for each side, the ratio of their volumes being as nearly 3 to 1 as the employment of convenient measurements will allow. They are cast in one piece with the cylindrical valve chamber and the saddle, the cylin-

Fig. 113

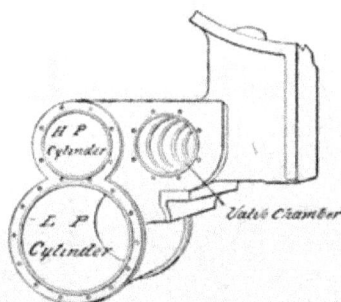

Baldwin Four-Cylinder Compound.
Cylinder Arrangement

ders being placed one directly above the other and as close together as they can be with adequate walls between them. Figs. 113 and 114 show the proximity of the two cylinders, while in Fig. 115, which shows the arrangement of the cylinders in relation to the valve, the actual construction is distorted for illustrative purposes.

The valve used to distribute the steam to the cylinders is of the piston type, working in a cylindrical steam chest located in the saddle of the cylinder casting as close to the cylinders as possible and between them and the smoke-box, as shown in the figures. This chest, having steam passages cast larger than required, is bored out enough larger than the diameter of the piston valve to permit the use of a hard cast iron bushing. Fig. 122 shows this bushing and one method of forcing it into place so that steam tight joints will be had between all ports; it

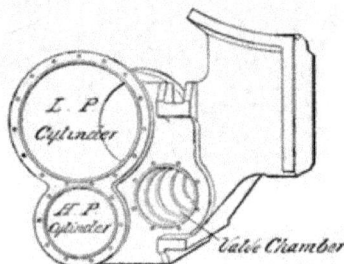

Fig. 114.

Baldwin Four-Cylinder Compound.
Cylinder Arrangement

also shows the narrow bridges across the steam ports which prevent the eight packing rings of the valve (shown in Fig. 116) from entering the ports. These cast iron packing rings form the edges of the valve.

The valve is of the piston type—double, and hollow between the two inside pistons—but having two solid ends, as shown by Fig. 116, and controls the admission and exhaust of both cylinders. The exhaust steam from the high-pressure cylinder becomes the supply steam for the low-pressure cylinder and is transmitted from one side of the high-pressure cylinder to the opposite

Fig. 115.

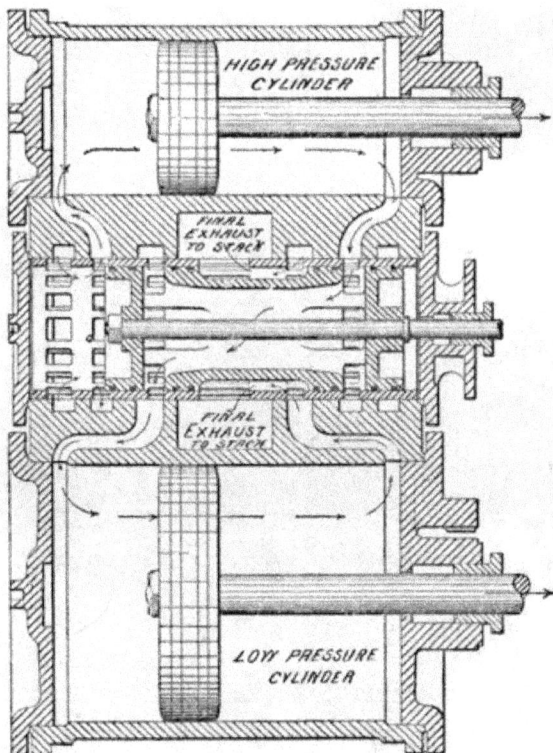

Baldwin Four-Cylinder Compound.
Steam Distribution with Piston Valve

side of the low-pressure cylinder through the hollow portion of the valve, as indicated by arrows, Fig. 115. The supply steam for the high-pressure cylinder enters the steam chest at both ends, thus balancing the valve with the exception of the area of the valve-stem at the back end.

The more common slide valve action being so much better understood by the average railroad man than the piston valve, I will liken this four-piston valve to one slide valve within another having external admission and internal exhaust in both cases.   Thus it will be seen that the outside

Fig. 116.

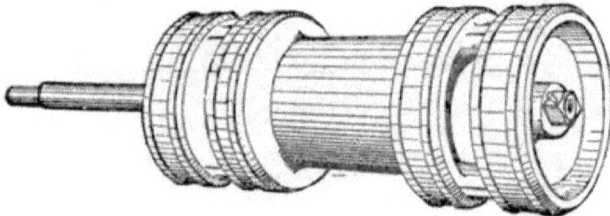

Baldwin Four-Cylinder Compound.
Piston Valve.

edges of the two outer pistons govern admission and their inside edges the exhaust of high-pressure cylinder, while the two inner pistons similarly regulate the flow of steam to and from the low-pressure cylinder, all of which will be evident by a reference to the arrows in Fig. 115.

Where the front rails of the frame are single bars, the high-pressure cylinder is usually put on top, as shown in Fig. 113, and in that event, with the usual rocker-arm, indirect valve motion is used.* When the low-pressure cylinder is put above (Fig. 114) on account of the double front rails of the frame, they also prevent the use of the rocker-shaft and box and the valve motion is then termed direct-acting, which necessitates a different location of the eccentrics on the axle.*

Engineers and those employed in shops and round-houses for setting valves and eccentrics should thoroughly understand the difference between the position of the eccentrics with relation

Fig. 117

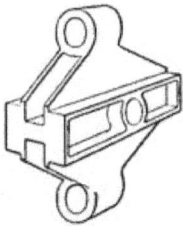

Baldwin Four-Cylinder Compound
Cross-Head

Fig. 118

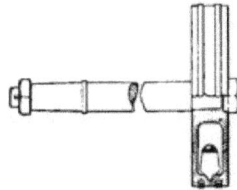

Baldwin Four-Cylinder Compound
Hollow Steel Piston

to the crank-pins for direct and indirect valve motion, as given fully elsewhere in the Manual, and further brought out in the Catechism on Accidents to Baldwin Four-Cylinder Compounds hereinafter contained.

The style of crosshead is shown in Fig. 117. It is made of cast steel, to insure the greatest strength with a minimum weight, the wearing surface being lined with tin. The piston, shown

---

*Direct and indirect valve motion will be found fully illustrated and explained in the earlier chapters of the Manual.

in Fig. 118, is also preferably made with cast steel heads, the object in both cases being to reduce the weight of the reciprocating parts to a minimum.

It is obvious, that, in starting these locomotives from a state of rest with heavy trains, it is necessary to obtain a greater power than that exerted by the high-pressure piston alone, for there would

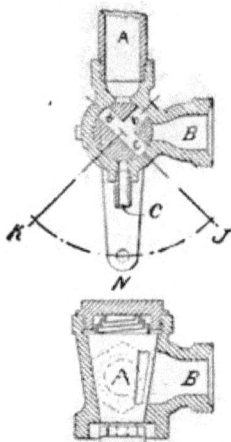

Fig.119

Baldwin Four-Cylinder Compound.
Combined Starting Valve and Drip Cock.

be no pressure on the low-pressure piston until the high-pressure cylinder had made one exhaust; hence it is necessary to admit steam to the low-pressure as well as the high-pressure cylinders. This is accomplished by the use of the starting valve (Fig. 119).* This is simply a plug-cock which is opened by the engineer by means of suitable levers from the cab, to admit steam from one end of the high-pressure cylinder to the other, and thence, as if it were the ordinary high-pressure exhaust, into the low-pressure cylinder. This same valve acts as a cylinder cock for both ends of the high-pressure cylinder and is operated by the same lever that actuates the ordinary cylinder cocks, which are in this case on the low-pressure cylinder, thus making, probably, the most simple starting device used on any compound locomotive and one not easily deranged. The operation of the starting valve in conjunction with the cylinder cocks is clearly shown in Fig. 120. The starting valve should be kept closed (position N) as much as possible, as its indiscriminate use reduces the economy and makes the locomotive "logy."†

Air valves, to prevent a vacuum, are placed in the steam passages of the high-pressure cylinder,

---

*This is sometimes called the "By-Pass" valve, as it connects the two sides of the high-pressure piston, but for an entirely different purpose than that to which the by-pass valves are put in connection with the low-pressure cylinder as described hereinafter under the Richmond and the Rogers compound, and for that reason I have not called it a "by-pass" valve. Two earlier forms of starting valves have been used with Vauclain compounds, but, inasmuch as they have been superseded by this form of valve, it is not deemed necessary to illustrate and describe them herein.

†An engine which should be capable of high speed but is not, and in which the pressures work against themselves in the cylinders, is said to be "logy."

a practice now generally followed on all locomotives, either simple or compound.   Additional air
valves, marked C and C^1 in Fig. 120, are placed
in connection with the ports in the valve chamber leading to the low-pressure cylinders.   Air
valves of somewhat different shape have been
described and shown in detail heretofore in the
Manual.

Water relief valves W W, Figs. 120 and 121,
which are nothing more nor less than pop valves,
are applied to the low-pressure cylinders and attached to the front and back cylinder heads to relieve excessive pressure of any kind, steam or
water.   The spring in the water relief valves on
these engines is made to carry a pressure enough
greater than the boiler pressure to prevent their
discharging steam and water ordinarily in starting the engine simple.

In all other respects the locomotive is the same
as the ordinary single expansion locomotive.

*Operation of the Baldwin Four-Cylinder Com-
pound.*—When starting the locomotive, the engineer should, ordinarily, pull the cylinder cock
lever way back and thus open the cylinder cocks
in order to relieve the cylinders of condensation,
and, as the starting valve is opened by the same
movement, steam is thus admitted to the low-
pressure cylinder and the locomotive started
quickly and freely.

In case the locomotive is at a platform of a
crowded station, or in any other place where it is

undesirable to open the cylinder cocks, the engineer should move the starting lever in the opposite direction from that usually given it,

Fig.120.

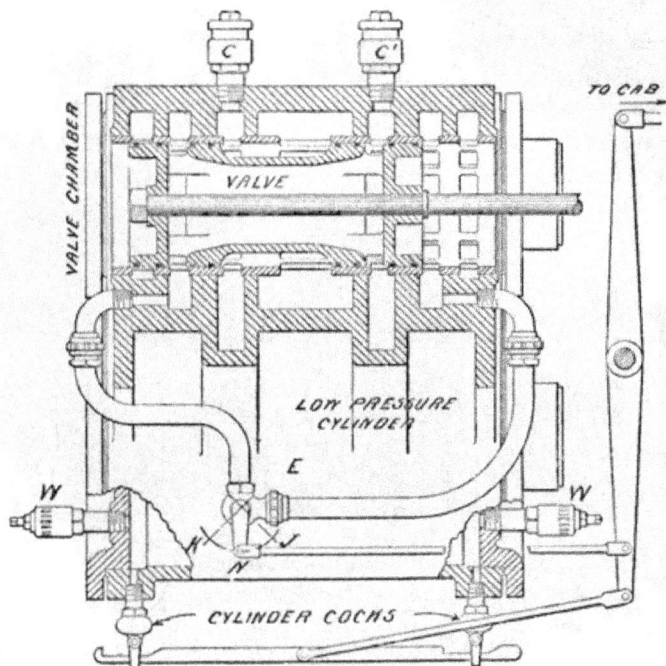

Baldwin Four-Cylinder Compound.
Showing operation of Starting Valve and
Cylinder Cocks.

placing the starting valve handle in position *J*, Figs. 119 and 120; that is, he should push forward the lever in the cab, thus allowing steam to pass

through the starting valve without opening either
the low-pressure cylinder cocks or the drip *C* of
the starting valves.   By further reference to Fig.
119, it will be seen how, when the handle is in position *K*, ports *A*, *B*, and drip *C* are all connected
by the ports *a*, *b*, and *c* of the plug; but if the handle
is in the opposite position *J*, ports *A* and *B* only
are connected, as *b* is now at *a* and *c* is opposite
B; in its central position *N* (normal position for
compound working), it will be seen that all ports
are closed as in the figure.

Fig. 121.

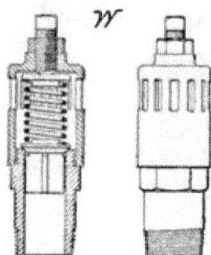

Baldwin Four-Cylinder Compound
Cylinder Relief Valve

After a few revolutions have been made and
the cylinders are free from water caused by condensation or priming, the engineer should move
the cylinder cock lever into the central position, *N*,
causing the engine to work compound entirely.
This should be done before the reverse lever is
disturbed from its full gear position.

Ordinarily, the reverse lever should not be
"hooked up," thereby shortening the travel of

the valve, until after the cylinder cock lever has been placed in its central position, but it is often necessary to open the cylinder cocks when at full speed to allow water caused by priming or foaming to escape from the cylinders, and in such cases no disadvantage is experienced, and the reverse lever need not be disturbed.

The starting device is simply designed for use in the starting of the train and should not be used at any other time unless there is imminent danger of stalling and the lever has been previously dropped to full gear. In other respects,

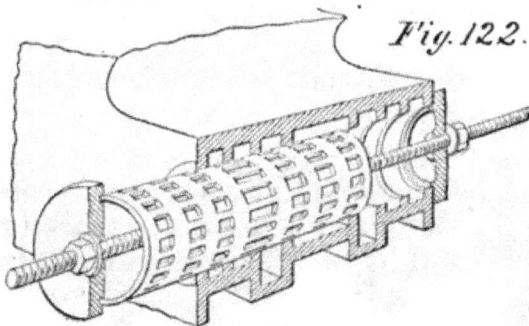

Fig. 122.

Baldwin Four-Cylinder Compound Valve Bushing, Showing Method of Pressing in

aside from these here noted, the rules governing the operation of compound locomotives in general should be clearly understood by any engineer who is liable to be called upon to run a compound locomotive of this or other design.

REPAIRS.

The builders of the Vauclain four-cylinder compound claim an advantage in it over the

two-cylinder or " cross-compound " locomotive
in simplicity of parts, there being no intercepting valve,* and a similarity to all the parts of a
single-expansion locomotive.   Thus its repairs
will be similar to those of simple locomotives.
To carry out this simplicity of parts, the piston
rods of the high and low-pressure cylinders are
of the same diameter and designed strong enough
to withstand the severest strains of service.

The packing rings in the valves are easily
replaced and the valve chest bushing can be
cheaply and easily renewed.   In extracting old
bushings it is best to split them between the.
ports with a narrow chisel.   The new bushings
can be pressed in by some such handy device as
that shown in Fig. 122.

Accidents to Baldwin Four-Cylinder (*"Vauclain"*)
Compounds.—For all ordinary accidents, such as
broken main rod or pin, or a broken valve stem,
what should be done?   The same as for non-
compound or simple locomotives, as described
fully in the earlier chapters of the Manual.

With a low-pressure cylinder head knocked
out, would it be necessary to disconnect that
side ?   Not for a short distance.

In that event, how many exhausts would there
be during one revolution ?   There would be three

---

*The intercepting valve is the valve which prevents the live
steam which is admitted from the boiler to the low-pressure cylinder at certain times, from passing through the receiver to the
high-pressure cylinder where it would produce back pressure on
the piston.

in the stack and one through the open cylinder head and the latter exhaust might obstruct the engineer's view, if on his side, and render the procedure inadvisable.

With the Vauclain Compound, at what position of the reverse lever is work of the two cylinders most nearly equalized? At a cut-off of about one-half the stroke in the high-pressure cylinder.

When is the work most unequal and the strains on the crosshead consequently the greatest? In starting with the engine working simple, as then the high-pressure piston is nearly balanced by live steam on both sides and the low-pressure cylinder obtains approximately boiler pressure.

What results would be likely should the rigging of the cylinder cocks and starting valve become bent or disconnected? Should one starting valve fail to properly close, the exhausts would be of unequal intensity. If one of them failed to open when required in starting, the engine would be weak on that side as it would have to start compound, that is with steam for the first stroke in the small high-pressure cylinder only.

In this latter event, when would the first exhaust from that side take place? Not until the completion of the return stroke.

If the cylinder cocks open and close with the same rigging as the by-pass valve, why would not the engineer know thereby that the by-pass valve was in position desired? From the previous description of this rigging, shown in Figs. 119 and 120, it should be remembered that the cab lever pushed clear ahead opens the by-pass valve, but not its drip nor the cylinder cocks.

Before altering the valve motion, what else should be examined if the exhausts were of unequal intensity? Examine for broken packing rings in the piston valve or the low-pressure cylinder.*

In case a valve-stem broke off inside the chest or the valve itself broke, would it be certain of discovery at once, as with an ordinary slide valve? Possibly it would not. Instances have been cited where compound locomotives of this system have hauled passenger trains long distances with broken valve-stems and broken valves. The two ends of the valve being unbalanced by the area of the valve-stem (see Figs. 115 and 116) accounts for the first possibility, while live steam from the induction ports acting on each end of the valve would explain the case of an undetected broken valve.

How can it be found if the cylinder packing in the high-pressure cylinder is blowing? Put the engine on the quarter, block the wheels, and test as usual for leaky slide valve; then, with the starting valve closed (in compound position) and the low-pressure cylinder cocks blocked open,

---

*A case is cited by the builders where an engineer ran his locomotive two days without any piston head at all in one of the high-pressure cylinders, and even then could not tell what was the matter except that the intensity of the exhausts were unequal and the engine did not make good time. Machinists put to work to locate the trouble, found it to the great surprise of the engineer.

drop the reverse lever into full gear. Steam passing the high-pressure piston will appear at the open cylinder cock of the low-pressure cylinder, but at the opposite end that would be expected with a simple engine.

How can it be found if the packing in the low-pressure cylinder is blowing? Put the engine on the quarter and open the starting valve and cylinder cocks and look for any escape of steam from the low-pressure cylinder cock on the end that should be in exhaust, as with a simple engine.

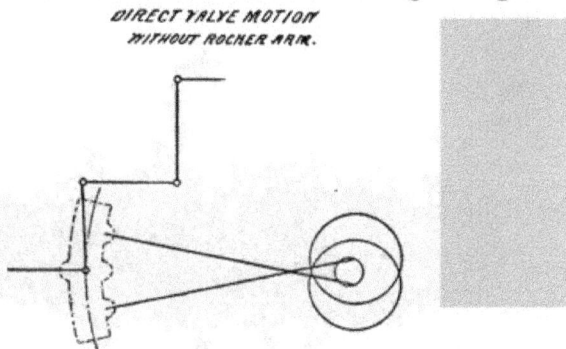

DIRECT VALVE MOTION
WITHOUT ROCKER ARM.

Fig. 123.

With the four-cylinder type, where the large low-pressure cylinder is placed on top, as in Fig. 114, and direct valve motion is employed. how should the eccentric rods on one side stand with the same side of the engine on the forward center? They should be crossed, as shown in skeleton Fig. 123. A slipped eccentric should be set the same as for similar valve motion on a simple engine, as fully described heretofore in the Manual under "Third Examination of Firemen."

THE BALDWIN TWO-CYLINDER COMPOUND.

The original Baldwin two-cylinder compound, built in the year 1892, was of the cross-compound receiver type and, after the first stroke or two.

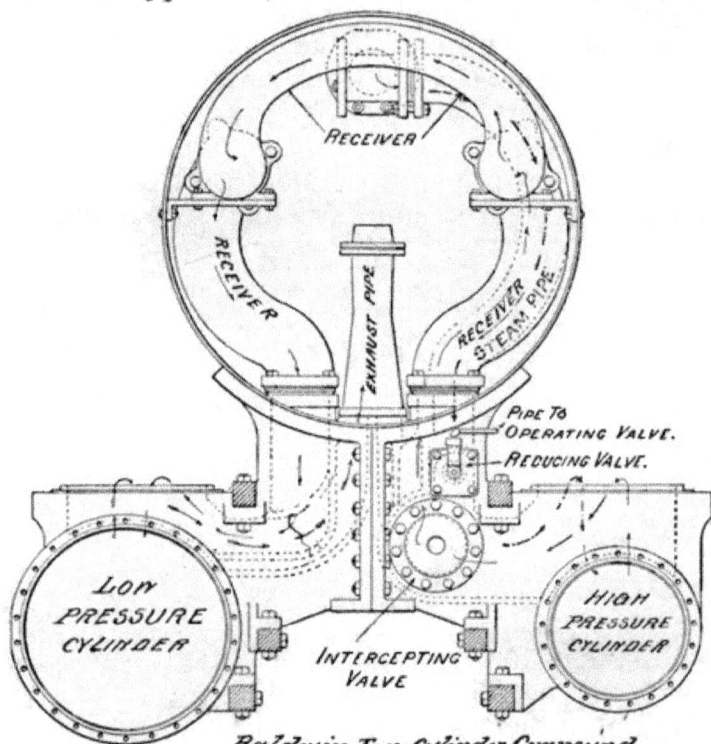

Baldwin Two-Cylinder Compound
Fig. 124

or as soon as the receiver had attained a pressure of 100 lbs., the engine automatically changed to compound and could not be operated otherwise. It belonged, therefore, to the automatic class of compounds. The reducing and the starting valve then employed were changed materially in the later design of the two-cylinder compound herewith illustrated and described.

Their later two-cylinder compound locomotives belong to the class of convertible compounds, as they can be operated either simple or compound for any length of time by the movement of a small valve in the cab, as shown by Fig. 128.

Fig. 124 shows a front view, giving the general arrangement of cylinders, steam, exhaust and receiver pipes in the front end, and the location of the intercepting and reducing valve in the saddle of the high-pressure cylinder. The low-pressure cylinder derives all its pressure from the receiver when running compound, as is usual in two-cylinder compounds.

The office of the intercepting valve is two-fold. It acts as an intercepting valve by opening and closing communication between the two cylinders, and also as a separate exhaust valve, by connecting the low-pressure cylinder with the exhaust to the stack. This it does by diverting the exhaust from the high-pressure cylinder either into the atmosphere, when working single-expansion, or into the receiver, when working compound, and is operated at the will of the engineer.

The office of the reducing valve is to admit live steam at a reduced pressure into the receiver and thence to the low-pressure cylinder, when the engine is working single-expansion, and also to close simultaneously with the changing of the intercepting valve to the position which causes

the engine to work compound, so that the receiver
will obtain no live steam from boiler when taking
the exhaust from high-pressure cylinder. The
performance of the first above-mentioned func-
tion—that of reducing the pressure of live steam
delivered to the receiver—is necessary in order

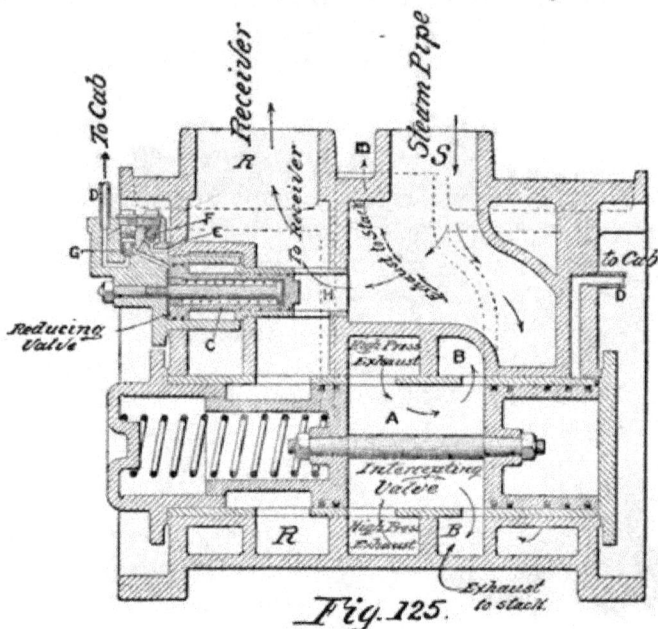

Fig. 125.
Baldwin Two-Cylinder Compound
Engine Working Simple

that the total pressure on the large low-pressure
piston shall not be greater than that on the high-
pressure piston, and thus the low-pressure side
kept from jerking the train and producing unequal
strains on the two sides of the locomotive when
working as a simple engine.

Operation of the intercepting and reduction valves.—In Figs. 125 and 126 the intercepting valve is marked *A* and the reducing valve *C*. It will be seen that they are both cylindrical in

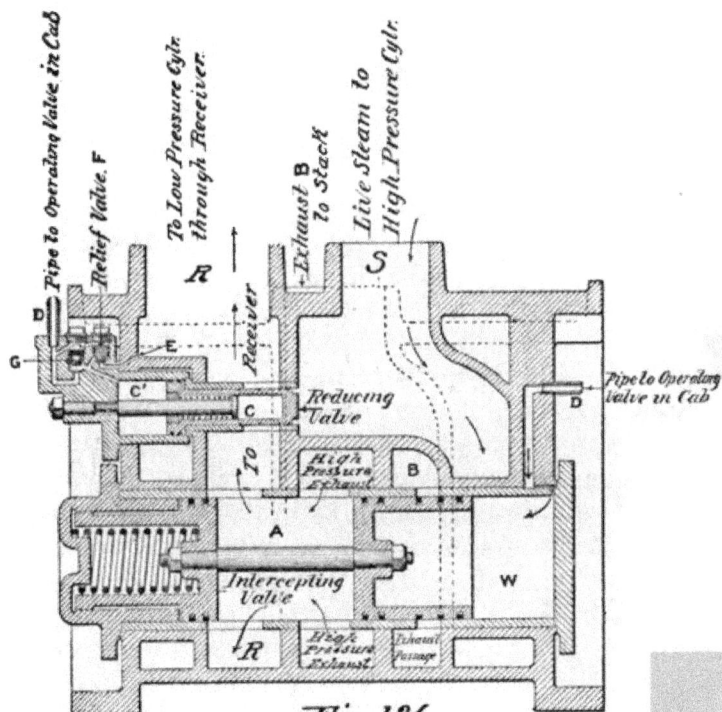

Fig. 126

Baldwin Two-Cylinder Compound Engine Working Compound.

form, are placed in bushings having suitable ports, and that coil springs hold them in their normal positions when no pressure is acting against them to overcome these springs.

In the cab of the locomotive is placed an operating valve, shown in Figs. 127 and 128, having two positions, marked "SIMPLE" and "COMPOUND." Through this operating valve a pressure of air or live steam is admitted to one side of the reducing and the intercepting valves through two pipes marked *DD*, and, acting against the right end of valve *A* and against the left end of valve *C*, moves both from their normal positions shown in Fig. 125 to those of Fig 126.

The reducing valve *C*, when it is not closed permanently by live steam from the operating pipe *D*, is automatically closed when the pressure in the receiver *R* is great enough to produce as much power in the large low-pressure cylinder as is obtained in the smaller high-pressure cylinder. For this purpose steam from the receiver *R* can pass through a port *E*, raising the poppet valve *F* (which remains open as long as the engine is not working compound) and bears upon the larger end of the reducing valve *C*, causing it to move to the right and close the live steam passage *H* (shown in Fig. 125) leading to the receiver *R*, whenever the receiver pressure becomes excessive. Thus it will be seen that when the engine is working simple there must be a close balance between the left-hand larger end of the reducing valve, being acted upon by receiver pressure, and the right-hand smaller end of the reducing valve, being acted upon by live steam from the main steam pipe *S*. In this way is the receiver pressure kept as much lower than the boiler pressure as the large end of the reducing valve is greater than the small end. This proportion is relative to the respective sizes of the high and the low-pressure cylinders and hence equal cylinder power will be given both sides of the engine in working simple. When the engine is standing, the lever of the small operating valve, Figs. 127 and 128, in the cab should be placed at position marked "SIMPLE," and the valves are then in position for the engine to work as a single-expansion locomotive, as the

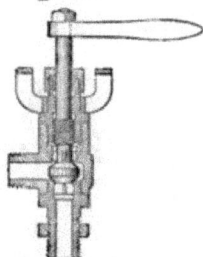

Fig. 127.

Dial

Fig. 128

Operating Valve
Baldwin Two Cylinder Compound.

steam pressure is relieved through this cab valve from the large end of the reducing valve and the right-hand end of the intercepting valve, allowing these valves to assume (by the action of their springs) their respective positions shown in Fig. 125. The arrows in this figure illustrate clearly how the steam can pass from the high-pressure exhaust through the intercepting valve A to the independent exhaust B leading to the stack (see dotted lines and arrows). At the same time the passage of live steam to the receiver—from which the low-pressure cylinder receives its supply—takes place through ports II, as shown by other arrows. The receiver pressure is governed by the automatic action of the reducing valve, as previously explained.

Thus the engine can be used as a single-expansion locomotive in making up and starting trains, and then, at the will of the engineer, the operating valve, Figs. 127 and 128, in the cab can be moved to the position marked "COMPOUND." This will admit live steam through the two supply pipes D, thence to the cylinders marked W and C", Fig. 126, changing the intercepting and the reducing valves quickly, and, as the ports are small, noiselessly, to the position shown in the latter figure. With the intercepting valve in this position it will be seen that the independent exhaust B is closed and steam from the high-pressure exhaust must follow the course of the arrows to the receiver, passing around the small reducing valve bushing and its valve C which is kept closed by the live steam from pipe D.

At any time the engineer may desire to increase the power of the engine as, for instance, when in danger of stalling, by moving the lever of the operating valve in the cab to position marked "SIMPLE" the engine is again changed at once to a single-expansion locomotive.

Accidents to Baldwin Two-Cylinder Compounds. —With one side disabled, what should be done in order to safely run the engine in ? Disconnect the disabled side, as advised for simple engines, place the intercepting valve in position for working simple so as to open the separate exhaust port, and run in with one side.

Should the small pipes DD leading to the reducing valve C and the intercepting valve piston be broken off, how could the engine be worked? With single-expansion only, unless the back head of the separate exhaust chamber W were removed and the piston blocked in the position shown in Fig. 126; then the engine would become an automatic compound, that is, would start simple but automatically go to compound after a revolution or so.

What would it be advisable to do in case of a broken reducing valve? Use very light throttle at slow speeds, or run with a reduced boiler pressure.

Should the small valves *F* and *G* be frequently inspected and cleaned? Yes. These valves and the reducing and the intercepting valves become gummed by the injudicious use of cylinder oil on the low-pressure side.

THE SCHENECTADY COMPOUND.

Locomotives built by the Schenectady Locomotive Works are oftentimes styled by the older railway men as "McQueen" engines, although the name of the builders has been as at present for many years.

These builders have constructed many compound locomotives, and, including the original valve design, have employed three styles of compound mechanisms, but all engines built have been of the two-cylinder variety of compounds.

Original Schenectady Type.—The original design by their then superintendent, Mr. A. J. Pitkin, consisted of an intercepting valve and a reducing valve. The stem of the intercepting valve was connected by levers to an index in the cab, which showed its position to the engineer. These engines belonged to the class of automatic compounds.

In starting the engine, a small pipe from the boiler through a reducing valve supplied steam to the low-pressure cylinder at a reduced pressure. When the receiver had accumulated sufficient pressure by the exhaust into it from the high-pressure cylinder, the intercepting valve would automatically be thrown to its normal position for working compound; then the supply of live steam to the low-pressure cylinder was cut off and the receiver pressure admitted, and thus the engine worked compound.

The following modification of this valve arrangement was afterwards made by Mr. Pitkin and applied to many locomotives by the Schenectady Locomotive Works.

Design of 1892.—With this construction of 1892, the opening of the throttle admits live steam at the same time to both the high and the low-pressure cylinders, closes the intercepting valve and allows the engine to start with its full power as a simple engine. After a few strokes the receiver pressure automatically opens the intercepting valve and cuts off the passage of live steam to the low-pressure cylinder and the engine works compound. The special valves are located in and behind the saddle on the low-pressure side and are operated automatically and beyond the will of the engineer. Fig. 129 shows the general appearance of that portion of the intercepting valve projecting back of the saddle; Figs. 130 and 131 show the valves and pistons removed from their encasing chambers. Upon opening the throttle, a small connection from the

steam pipe admits live steam through suitable
valves to an actuating piston, the movement of
which opens a poppet valve, supplying live
steam to the low-pressure cylinder, and also places
the intercepting valve so as to close connection
between the receiver and the low-pressure steam

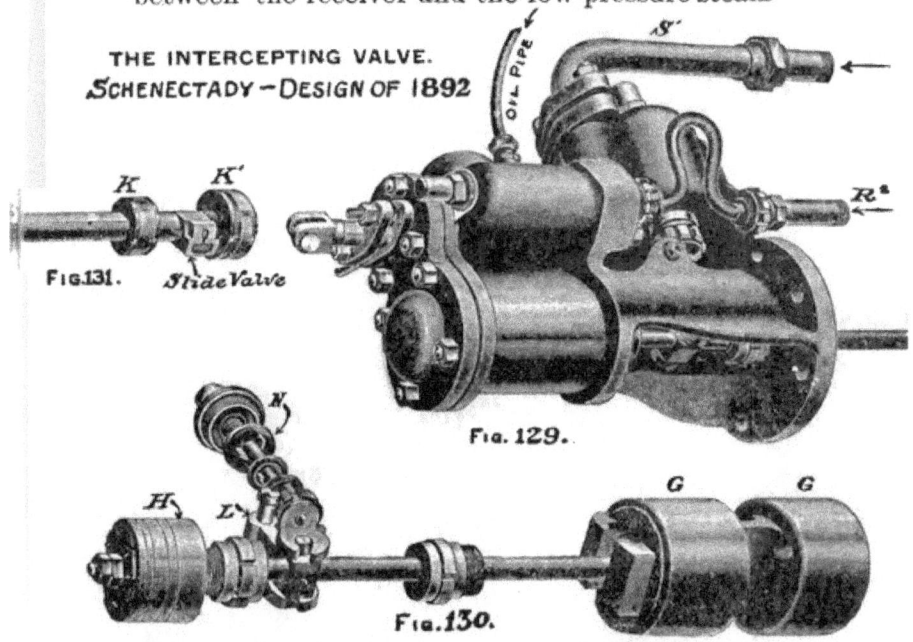

THE INTERCEPTING VALVE.
SCHENECTADY — DESIGN OF 1892

FIG.131.   *Slide Valve*

FIG. 129.

FIG. 130.

chest. Thus the low-pressure cylinder exhausts
to the atmosphere and the high-pressure cylinder
into a closed receiver. Sufficient pressure will
accumulate in the receiver after a few strokes to
move the small valves, thereby moving the actuating piston and with it the intercepting valve, to

such position as will close off live steam to the low-pressure cylinder, and instead admit the receiver pressure, thus working the engine compound.

For the benefit of those interested in the details of this device, a more thorough description of the accompanying figures follows:

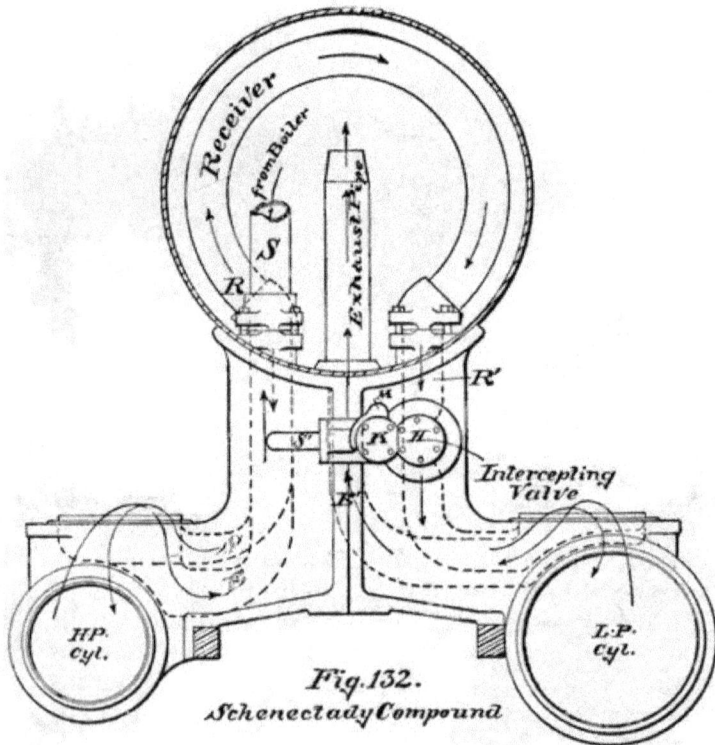

Fig. 132.
Schenectady Compound

The front view, Fig. 132, shows the general arrangement of cylinders, steam passages, and the intercepting valve. Figs. 133 and 134 both show the same horizontal section through the saddles and show the intercepting valve and the actuating valves, Fig. 133 showing them in position for working compound, and Fig. 134 for starting. Fig. 135 gives a vertical section, better showing the passages between the receiver and the low-pressure steam chest, which passages are opened and closed by the double pistons *GG* which form the intercepting valve. Of the remaining figures, 136 and 137 show details of the regulating valve, and Fig. 138 an end view of the intercepting pistons *GG*. The arrows in Figs. 133 and 134 indicate the direction of the steam in passing through the apparatus.

Fig. 132 shows a smoke-box mounted on saddles connected with the high and low-pressure cylinders located on opposite sides of the engine and having the necessary admission and exhaust ports. The exhaust port of the high-pressure cylinder is connected by a passage *E* (see dotted lines in Fig. 132 and full section of port in Figs. 133 and 134) with the receiver at *R*, Fig. 132. The other end of the receiver connects with the inlet passage R^1 (shown also in Fig. 135) leading to the low-pressure steam chest, and in this passage the intercepting valve *GG* is located and travels across it to open or close this passage.

The intercepting valve and the mechanism for operating it are mounted on the saddle of low-pressure cylinder, as before stated, while the live steam pipe *S* and the high-pressure exhaust passage *E* are situated in the high-pressure saddle. The low-pressure exhaust passage E^1 is formed

by the two saddles being bolted together, see
Figs. 133 and 134.

The intercepting valve consists of two pistons
GG (having several small holes *g* through them
in order to balance them, see Figs. 135 and 138)
mounted at one end of a long piston rod, which

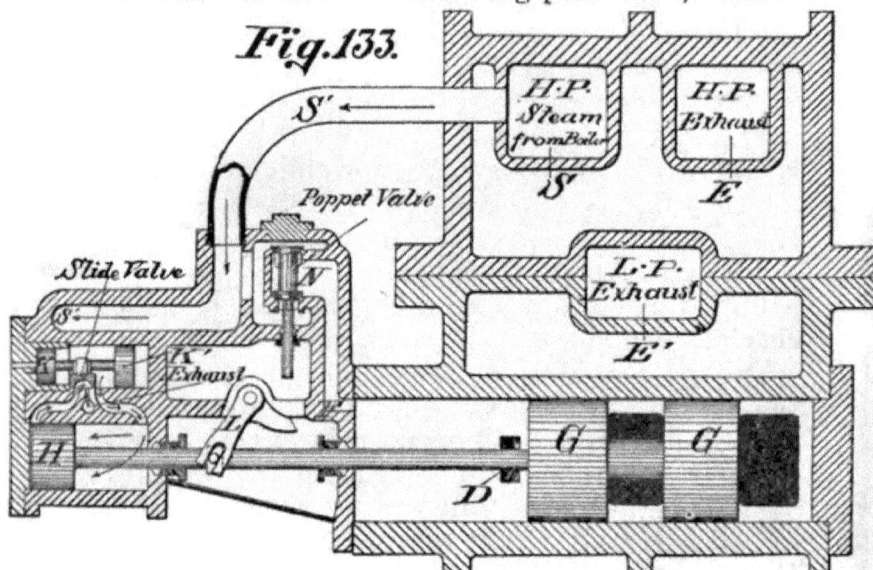

Fig.133.

Intercepting Valve Open
Engine Working Compound

moves to and fro in a cylinder having four openings. The two large openings shown lead from
the receiver to the low-pressure steam chest (see
Fig. 135) and are closed by the two intercepting
pistons *GG* when the engine is to be started, so
that live steam may be admitted to the low-

pressure cylinder without producing a back
pressure in the high-pressure cylinder through
the receiver. Of the two remaining openings in
the valve cylinder, port *D* leads to the low-pressure
steam chest and port *F* admits steam from the
boiler when the apparatus stands in position as
shown in Fig. 134.

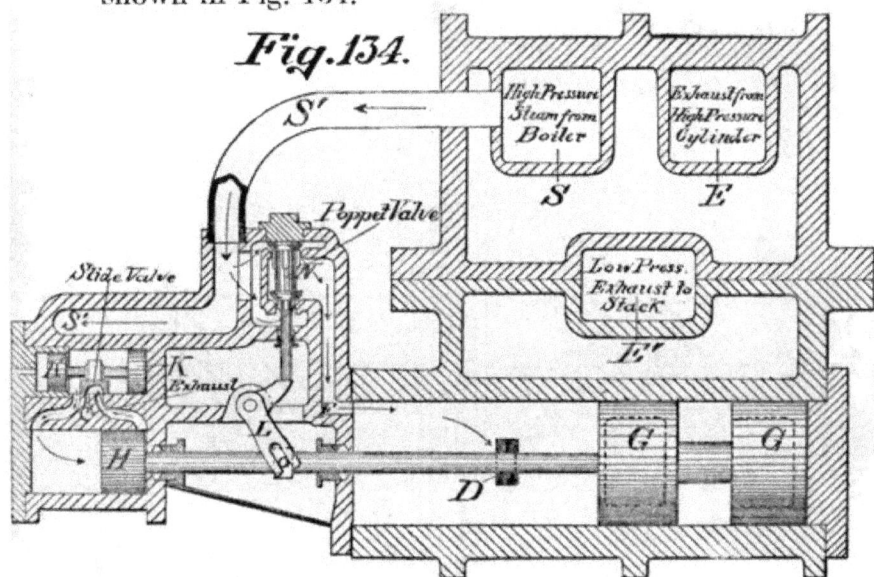

Fig.134.

Intercepting Valve Closed
Engine Working Simple

The back end of the intercepting piston-rod
passes through suitable stuffing boxes to a small
cylinder provided with a piston *H*, which actuates
the intercepting valve. This cylinder has a small
steam chest, slide valve and admission and exhaust

ports so similar to those of an ordinary locomotive cylinder, that its operation will be made plain
by referring to the Figs. 133 and 134, if the movements of its slide valve are explained.   This

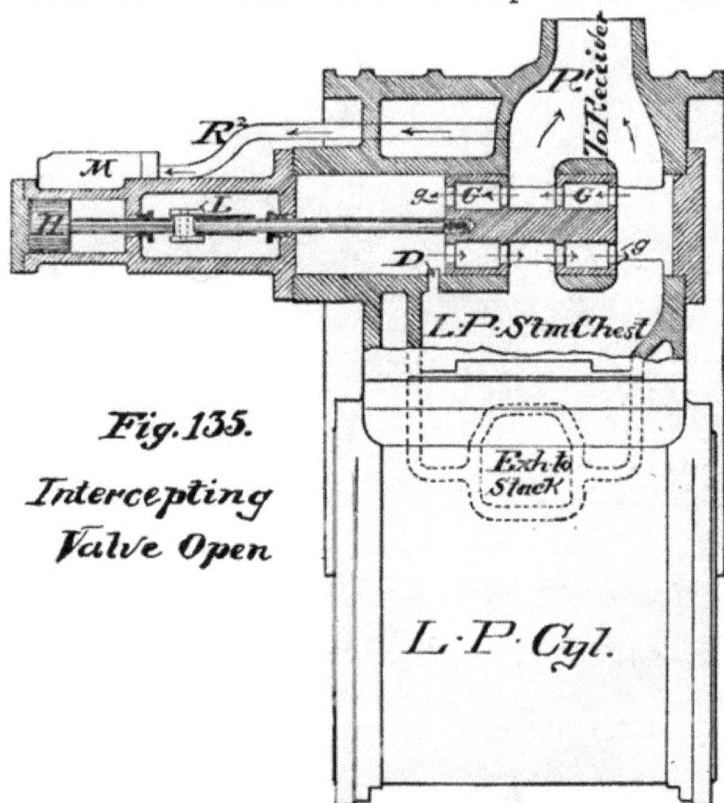

Fig. 135.

Intercepting
Valve Open

small slide valve, Fig. 131. is moved by a stem
connecting two pistons, K and K^1, of unequal
diameter in order to insure their movement in
the proper direction at the proper time.   The
actuation of these pistons, and with them the slide

valve, will be made clear by the figures. From Fig. 135 it will be seen that a small pipe R^2 leads from the receiver connection R^1 to this valve mechanism, and from Figs. 133 and 134, that a pipe S^1 comes from the live steam passage in the saddle and has a small port leading to the actuating valves as well as to the poppet valve N. These live steam and receiver connections come to opposite sides of a small piston valve M (Fig. 137), which is called the "regulating valve" and travels across two ports leading to the slide valve beneath it, as shown.

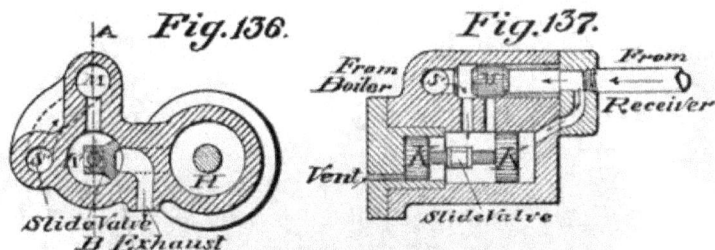

The remainder of this mechanism consists of a balanced poppet valve N, which, when open, admits live steam from pipe S^1 through the intercepting valve to the low-pressure cylinder in starting. This poppet valve N has a projecting stem on the lower side and is opened and allowed to close by a rocker-arm or bell-crank L, its two positions being shown in Figs. 133 and 134, respectively.

The operation of the apparatus is as follows: The normal position of the parts when the engine is working compound is shown in Fig. 133, in which position steam for the low-pressure

cylinder comes entirely from the high-pressure
exhaust through the receiver.   To start the train,
the engine throttle is opened as usual.   This permits steam to pass to the high-pressure side and
also through the pipe S^1 (Figs. 133 and 136) to
the left side of piston valve M (Fig. 137) and
down through the adjacent port (as indicated
by arrows) to the slide valve chamber, there acting between the two pistons K and K^1 (Figs.
131, 133 and 137).   The right-hand piston, being
the larger, causes a movement of the slide valve
from its position shown in Fig. 133 to that shown
in Fig. 134, thereby uncovering the steam port
to the left of piston II, which it forces with the intercepting valve GG to the right. In this position,

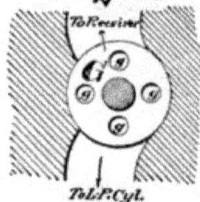

Fig.138.   as shown in Fig. 134, the receiver
openings are closed by the pistons
GG and the poppet valve N has
been opened by the bell-crank L,
thus admitting live steam through
the intercepting valve cylinder and
port D to the low-pressure steam
chest, as indicated by the arrows.   Hence it is
possible to obtain the full pressure of live steam
in the low-pressure cylinder in starting.

After one or two revolutions the pressure in the
receiver, passing down through the small connecting port to the right of the larger piston K^1 (Fig.
137) overbalances the pressures between the pistons, thus moving the slide valve to the left, the
position shown in Figs. 133 and 137. According to
the ordinary action of a slide valve this reverses
the pressures on the actuating piston II, forcing it

to the left and opening the intercepting valve.
This return movement of the actuating piston *H*
detaches the bell-crank *L* from the poppet valve
N and allows the latter to close before the intercepting valve opens. After this the locomotive
works compound, the passage of steam being
through the high-pressure cylinder to the receiver
and thence through the intercepting valve and
low-pressure cylinder to the atmosphere, as previously described.

A difficulty met with in many of the earlier
forms of compound mechanism, and to which the
reader's attention was called at the beginning of
this chapter, namely, the accumulation of dan-
gerously high pressure in the receiver when running with the throttle closed, was overcome in
this device by an automatic action of the piston
valve *M* and the differential pistons *K* and *K'*
(Fig. 137), as follows: When the engine is using
steam the regulating valve *M* is always against
the right-hand seat, as shown, and this valve only
comes into use when running without working
steam, as down a long grade. In this case, if the
intercepting valve happened to be closed, the
action of the engine would cause air-pressure to
accumulate in a closed receiver as there would
then be no live steam available to cause the
actuating device to open the intercepting valve.
Hence it is arranged so that air-pressure in the
receiver will force the valve *M* to the left and
itself take the place of live steam by passing to
the slide valve chamber and down to the right
side of the actuating piston *H*, moving it to the

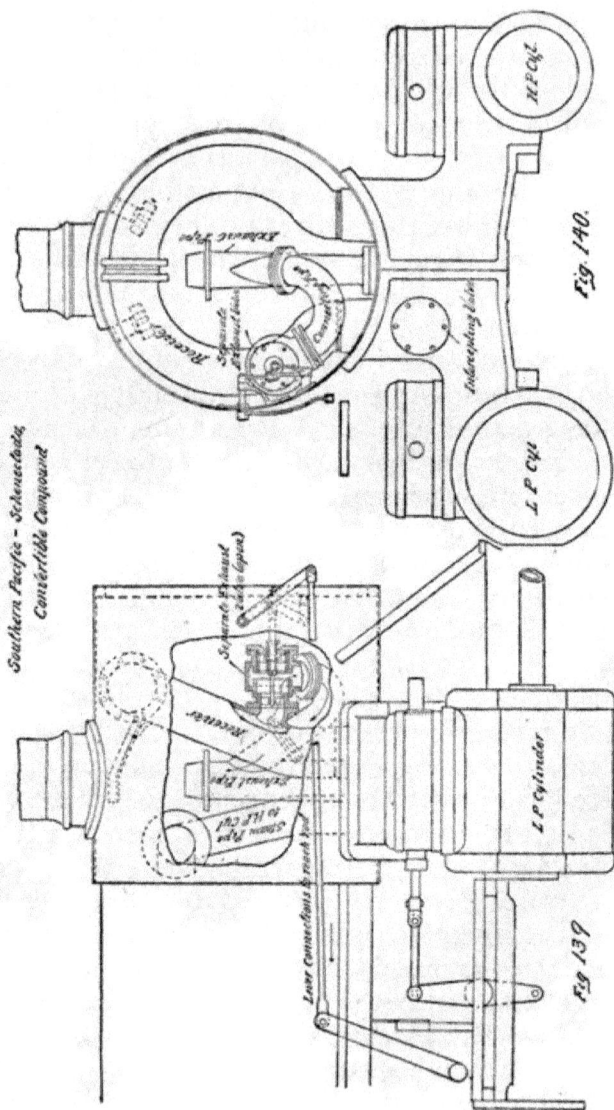

Southern Pacific – Schenectady,
Convertible Compound

Fig. 140.

Fig. 139

left and opening the intercepting valve, as shown in Fig. 133. Thus the small valve *M* acts as a safety valve, insuring the opening of the intercepting valve when live steam is not being used, and preventing the danger of excessive receiver pressure or the lifting of the high-pressure slide valve off its seat when the engine is running with steam shut off.

SCHENECTADY 1892 DESIGN, WITH SOUTHERN PACIFIC MODIFICATION.

To render it possible to run the engine "simple" for any desired period in starting, or to obtain a maximum power in case a train were stalling on a heavy grade, the Southern Pacific Co. in 1893 added to many of their Schenectady compounds of the 1892 design, a separate exhaust valve located in the smoke-box, as shown in Figs. 139 and 140. The reverse lever in the cab, when placed in either of its extreme positions, caused this valve to open and thereby connect the receiver directly with the main exhaust pipe, thus permitting the high-pressure cylinder to exhaust through the receiver directly to the atmosphere, as indicated by arrows in Fig. 139. As the receiver pressure was thus kept down it will be readily understood from the preceding description of the intercepting valve that the latter will remain in starting position as in Fig. 134, and hence the locomotive will work as a simple engine until such time as the engineer pulls the reverse lever higher up on the quadrant and thereby closes the separate exhaust valve. Then the

intercepting valve automatically assumes the
compound position, as in Fig. 133, for reasons
hereinbefore explained.

This modification of the two-cylinder or cross-
compound is of especial note inasmuch as it was
one of the first in this country which permitted
the working of the locomotive as a simple engine
for any desired length of time, at the will of the
engineer. Its results in practical operation were
to greatly reduce the jerking of trains in starting
(then a very serious objection to many compounds); it gave a greater maximum power at
critical periods, and was withal so eminently satisfactory that the reader will notice the majority
of the builders of two-cylinder compounds in this
country have embodied a separate exhaust valve
in their later designs.

SCHENECTADY COMPOUND—DESIGN OF 1896.

The valve arrangement designed in 1896 by
Messrs. A. J. Pitkin, Vice-President and General
Manager, and J. E. Sague, Mechanical Engineer of
the Schenectady Locomotive Works, and used as
their standard construction for two-cylinder compound locomotives, will be made clear by what
follows.

In general it may be said that this so-called
"intercepting valve" consists of four separate
parts, namely: (1) An intercepting valve proper,
which allows steam to pass to the low-pressure
cylinder from either the receiver or the boiler,
according to its position. (2) A reducing valve
allowing live steam at only a reduced pressure

to enter the low-pressure cylinder when working simple. (3) An independent or separate exhaust valve which, when open, vents the exhaust from the high-pressure cylinder to the atmosphere through the exhaust pipe and stack. (4) A small valve *K* inside of the separate exhaust valve, by the use of which the latter can be opened more easily and gradually.

Fig.141.   **Fig.142.**

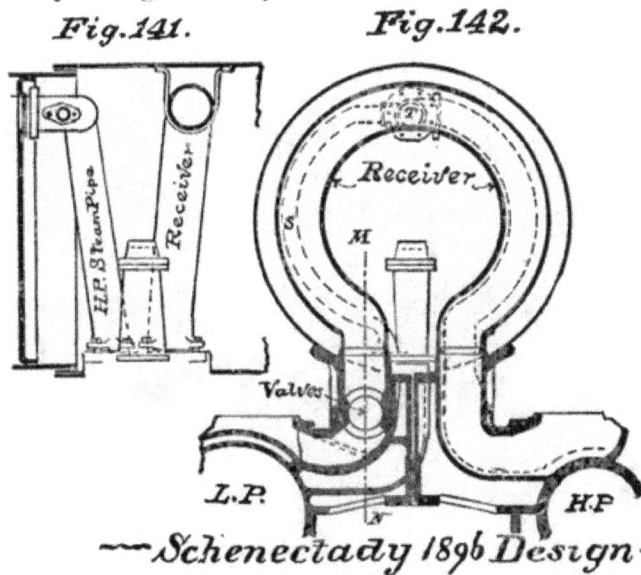

~Schenectady 1896 Design~

By the arrangement of these valves the engine can be started and run either compound or simple and can be changed from compound to simple, or the reverse, at the will of the engineer, with the throttle and the reverse lever in any position; the engineer has only to move a small three-way cock in the cab and the working of the engine changes very smoothly and without jerking the train.

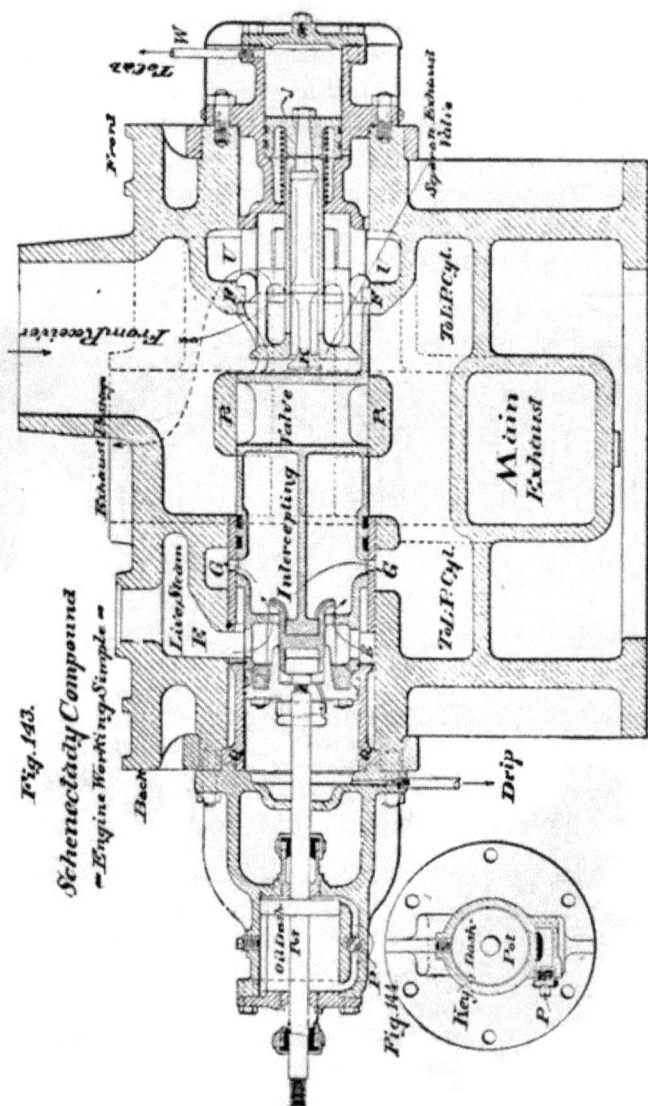

Fig. 143.

Schenectady Compound

— Engine Working Simple —

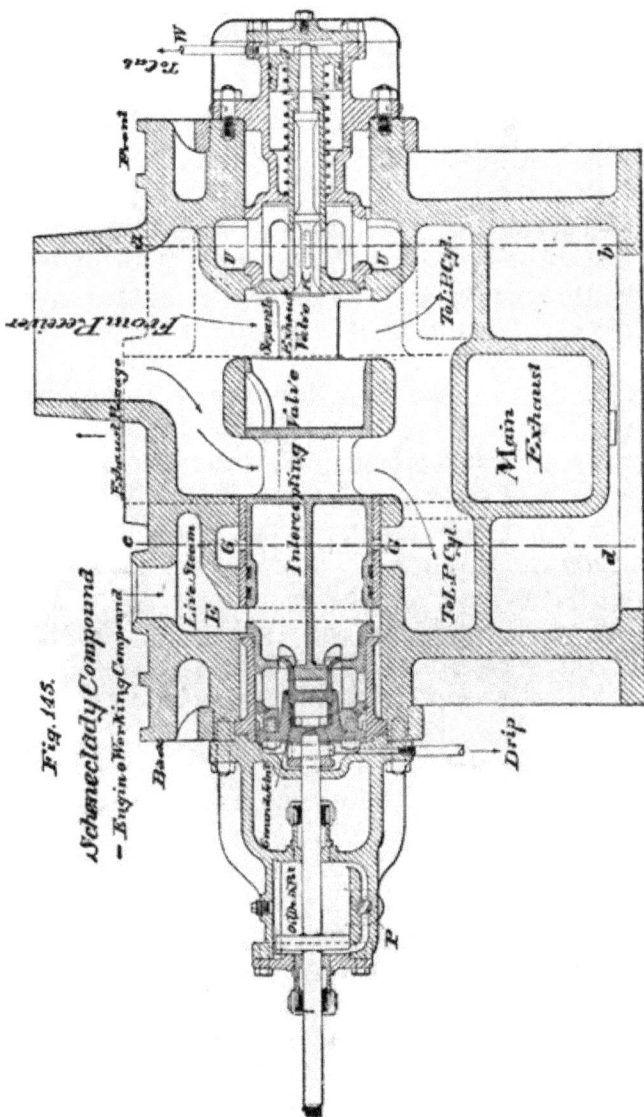

Fig. 145.

Schenectady Compound

— Engine Working Compound

Figs. 141 and 142 give sections of smoke arch and cylinder saddles and show the steam passages, the receiver and the location of the intercepting valve in the saddle of the low-pressure cylinder on the right-hand side of the engine.

It will be noticed by the dotted lines behind the receiver pipe that there are two steam pipes as in a simple engine, but the one (S) leading to the intercepting valve on the low-pressure side is much smaller than usual, as it will only be required for use at low speeds.

Fig. 143 shows a vertical section lengthwise through the low-pressure cylinder saddle and the intercepting valve (as if they were cut through at MN of Fig. 142) and shows the intercepting and the separate exhaust valves in the position taken when the engine is working simple and receiving live steam in both cylinders. Fig. 144 is a section through the dash-pot of Fig. 143.

Fig. 145 gives the same section as Fig. 143, but shows the intercepting and the separate exhaust valves in the position taken when the engine is working compound.

Figs. 146 and 147 show two sections crosswise of the intercepting valve at points indicated respectively by the lines *cd* and *ab* of Fig. 145. Section *cd* shows the passages G for admitting live steam into the low-pressure cylinder, and section *ab* shows the outlet passage U from the separate exhaust valve to the main exhaust pipe.

The part which each portion of the valve arrangement performs is as follows: The separate exhaust valve, when open, allows the steam to exhaust from the high-pressure cylinder to the atmosphere without going through the low-pressure cylinder, thus working the engine simple; when it is closed, the high-pressure exhaust must pass through the low-pressure cylinder, thus working the engine com-

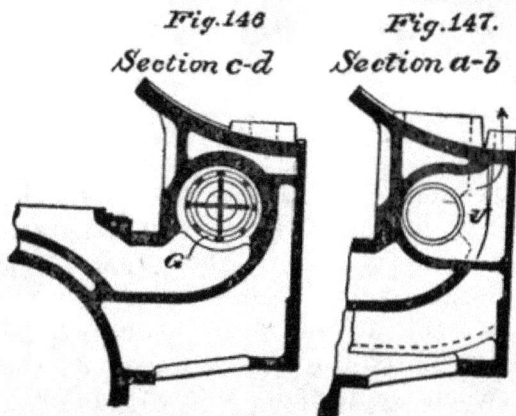

Fig.146   *Fig.147.*

Section c-d   *Section a-b*

-Intercepting Valve Passages-

pound. The intercepting valve closes the passage between the two cylinders when the separate exhaust valve is open, so that steam cannot go from the high-pressure cylinder to the low-pressure cylinder; thus doing away with back pressure on the high-pressure piston when the engine is working simple; it also admits live steam direct from the dry-pipe through the reducing valve to the low-pressure cylinder. When the separate exhaust valve closes, the intercepting valve automatically opens the pas-

sage between the two cylinders and cuts off the
supply of live steam from the dry pipe to the
low-pressure cylinder. The reducing valve works
only when the engine is working simple and
throttles the steam passing through it, so that
the pressure of steam going to the low-pressure
cylinder is about one-half (or, less, according to
the proportionate sizes of the two cylinders) of
that admitted from the boiler to the high-pressure cylinder.

The reducing valve is quite heavily cross-sectioned, while the long, intercepting valve appears next lighter, in order to render their out-
lines in Figs. 143 and 145 readily distinguishable.
Examining the two ends of the intercepting valve,
it will be seen that the left end, exposed to the
pressure of the atmosphere through the drip, is
only about three-fourths as large as the right
end (between the bridges $R R$, Fig. 143), exposed
to the receiver; hence, if the receiver has little
or no pressure, the boiler pressure on the
shoulder of the intercepting valve automatically
carries it to the right, as shown in Fig. 143. The
reducing valve is automatically opened because
of the difference in area of its two ends also. The
movement of each of these valves is cushioned
by dash-pots, as shown. The separate exhaust
valve is operated by the engineer by means of a
three-way cock in the cab. To open the separate exhaust valve the handle of the three-way
cock is thrown so as to admit a pressure of steam
or air through the pipe W against the piston J.
Pulling the handle back relieves the pressure

against piston *J* and the spring shuts the valve, as in Fig. 145. All the engineer has to do in connection with the operation of the valves is to pull the handle of the three-way cock in the cab one way or the other, according as he wishes the engine to run simple or compound. The engineer uses the handle under the following conditions:

To Start Simple.—Under ordinary conditions this is not necessary, but if the maximum power of the engine is needed to start a heavy train, the engineer pulls the handle of the three-way cock so as to admit pressure from the cab through pipe *W* against the piston *J*, Fig. 145. This will force piston *J* into the position shown in Fig. 143, opening the separate exhaust valve and holding it open. The engine throttle now being opened, live steam at boiler pressure enters the chamber *E* from the small steam pipe *S* before mentioned and forces the intercepting valve to the right against the seat *FF*, as shown in Fig. 143. The exhaust steam from the high-pressure cylinder now passes through the receiver and is exhausted through the separate exhaust valve to an annular chamber *U* connected with the main exhaust to the stack, as indicated by the arrow in Fig. 143. (See also Fig. 147.) Steam also enters the low-pressure cylinder from chamber *E* through the reducing valve and the annular ports *G* in the intercepting valve (See Figs. 143 and 146), and is exhausted in the usual way. The reducing valve prevents the full boiler pressure from reaching the low-pressure cylinder. As will be seen from

Figs. 143 and 145, the reducing valve is partly balanced by its smaller left end being open to the atmosphere through a small groove leading to the chamber having an open drip, and thus the boiler pressure acting on the unbalanced area throws the valve open—to the right. When the pressure in the intercepting valve cavity on the right of the reducing valve becomes high enough, it will throw the valve to the left, because it acts on the whole area of the valve; the result is that the steam is throttled to the proper pressure desired for the low-pressure cylinder.

To Work Compound.—Having started the train, when the engineer wishes to change the engine from simple working to compound, he pushes the handle of the three-way cock to its first position which, relieving the pressure on piston J through pipe W, allows the spring to act to the right and close the separate exhaust valve, as in Fig. 145. As soon as this valve is closed the pressure in the receiver, having no outlet, rises and presses the intercepting valve to the left against the pressure from chamber E, which acts only, as stated, upon the shoulder of the intercepting valve. The receiver pressure holds the intercepting valve to the left, as shown in Fig. 145, thereby closing the ports G and opening a free passage from the receiver to the low-pressure cylinder as indicated by the arrows, and the engine works compound. While working compound, which is the usual way of working the engine, both the reducing and the intercepting valves are held to the left against ground joint seats. This should prevent any steam which might leak by the packing rings from constantly escaping at the drip.

To Change from Compound to Simple.—With the engine running compound, if the engineer wishes to change to simple because of a very heavy grade, he has only to pull the three-way cock handle to the same position as for starting simple. Then piston *J* first opens the small valve *K* and then the separate exhaust valve. The small valve *K* relieves the pressure more gradually than if the larger valve were opened at once. As soon as the separate exhaust valve is opened the pressure in the receiver escapes through it and becomes so low that the intercepting valve is again forced to the right (as in Fig. 143) against its seat *F* by the steam pressure from chamber *E*, and the engine works simple as in starting.

To Start as an Automatic Compound.—If the separate exhaust valve is left closed, as in Fig. 145, the engine will start as an automatic compound when the throttle is opened, for the pressure from chamber *E* will force the intercepting valve to the right, as in Fig. 143, thus admitting live steam through the reducing valve and ports *G* to the low-pressure cylinder, while at the same time the high-pressure cylinder exhausts into a closed receiver for a few strokes. This pressure, accumulating in the receiver, will then automatically close the ports *G* by moving the intercepting valve to the left, as in Fig. 145, and the engine thereafter runs compound.

Accidents to Schenectady Compounds—the Automatic Compound of 1892. What should be done

in case of a break-down on the road, necessitating the disconnecting of the high-pressure
side? If but a short distance to go and a slow
speed would suffice, clamp the high-pressure
slide valve in center and permit the engine to
run by the admission of live steam through
the small pipe S' and the poppet valve N to the
high-pressure cylinder (Figs. 133 and 134). If
the intercepting valve is out of order, block the
poppet valve N open, that is, up. If it were required to run at considerable speed, this small
pipe S^1 would give insufficient supply, in which
case the high-pressure slide valve should be
blocked clear back (much farther than its ordinary travel carries it), so as to uncover the
exhaust port, thus admitting live steam direct
to the receiver. If the steam chest is large
enough to place the high-pressure valve as
described, and the intercepting valve is not deranged, the engine would run at full speed with
the low-pressure side. If out of order, the intercepting valve should be held open (in the position
as shown in Fig. 133) by clamping the stem between the stuffing boxes. In all cases the throttle
should be handled easily to prevent a too rapid
flow of boiler pressure to the large low-pressure
cylinder and the consequent liability of jerking
the train or causing damage to this cylinder.

What should be done if it becomes necessary
to take down the low-pressure side of the engine?
The engine could be moved a short distance with
the cylinder cocks open or the indicator plugs removed on the high-pressure side, but as most en-

gines of this class have either a large steam chest or an "Allen"[*] ported slide valve, the valve can be clamped back far enough to uncover the low-pressure exhaust port, and thus run at full speed. If this cannot be done, block both the low-pressure crosshead and valve clear back and unscrew the relief valves or take off the front cylinder head on that side to make an exhaust opening from the receiver. If the intercepting valve is out of order, it must be securely clamped open, as in Fig. 133, otherwise the opening between the receiver and the low-pressure steam chest would be closed.

In this last procedure, with the exhaust other than through the stack, would the engine steam with much of a train? No; but a limited amount of steam could be maintained by the use of the blower for creating draught.

What would be the effect of the removal of the slide valve on the disabled side? This would give a free port opening under all circumstances, but would generally consume too much time to be practicable.

What prevents the leakage of live steam into the receiver when the intercepting valve is closed, as in Fig. 134, there being no packing rings in the two pistons GG? The live steam pressure acts from below when starting, so as to hold these pistons tight against ports of the receiver. Fig. 135 illustrates this clearly, if the intercepting

---

*See illustrations in Part First of the Manual, showing the "Allen" slide valve.

valve were there shown closed, as live steam
would then be below pistons *GG*.

What would be the result if the wiper *L* would
strike the poppet valve *N* (Figs. 133 and 134)
before the intercepting valve pistons *GG* closed
their ports?   Live steam would blow through to
the receiver and produce a back pressure on the
high-pressure side.

How can this be prevented?   Pistons *GG* have
sufficient lap to allow of their closing before the
wiper *L* strikes the poppet valve *N*, and the
adjustable tappet on the intercepting valve stem
should be set so as to cause this.   If the tappet is
set too far back, valve *N* would not be opened at
all and, as a consequence, no live steam would be
admitted to the low-pressure cylinder in starting.

If the operating piston *H* should break, what
position would the intercepting valve probably
take?   On account of the unbalanced area of the
stem, it would probably move open to the left
as for compound working, Fig. 133.

*Accidents to Schenectady design of 1892, with
Southern Pacific Modifications.*—If it became
necessary to disconnect the high-pressure side of
the engine, what should be done?   The same as
with the 1892 Schenectady system.

Would there be any difference in case the low-
pressure side broke down?   Yes; disconnect the
broken side as usual (see instructions for simple
engines in Part First of the Manual) and run with
reverse lever in full gear, if for a short distance or
a low speed only is required.   If it is necessary to
run for a considerable distance at a good speed, it

would be advisable to disconnect the separate exhaust valve levers from their connection to the reach-rod and properly secure them in either extreme position, so as to hold the valve open. The engine can then be "hooked-up," that is, the reverse lever pulled up toward its central position, to correspond to the demands of the service.

Accidents to Schenectady Compounds—design of 1896.—What should be done in case the high-pressure side had to be disconnected? Ordinarily, open the separate exhaust valve * and do nothing different than with a simple engine; but to obtain greater speed than the supply of live steam to the low-pressure cylinder through its small steam pipe would permit, the high-pressure valve should be secured in such a position, if possible, as will uncover its exhaust port, thereby admitting live steam to the receiver and thence to the low-pressure cylinder. In this case leave the separate exhaust valve closed and handle the throttle easily so as not to cause constant opening of the safety valves on the low-pressure side.

What is necessary with the low-pressure side disconnected? Open the separate exhaust valve and allow the high-pressure cylinder to exhaust to the stack through its connection. While considerable train could thus be handled, it would not be done at anything but a slow speed, unless the low-pressure slide valve were placed so as to

* While not absolutely necessary to open the separate exhaust valve for this case, it is best to do so that there may be no accumulation of pressure in the receiver should the high-pressure valve leak.

uncover its exhaust port and the separate exhaust valve left closed.

What is done to prevent full boiler pressure from reaching the low-pressure cylinder in case the reducing valve becomes defective or broken? Pop or safety valves are placed on the chest and both heads of the low-pressure cylinder and they are set at about one hundred pounds, the highest pressure deemed advisable in so large a cylinder.

In case of a broken intercepting valve what precautions should be taken? Run the engine compound only and do not stop the engine with the low-pressure side on center.

Why must the oil dash-pot be kept filled with oil? The flow of oil from one side of the dash-pot piston to the other prevents sudden movements of and serious jars to the intercepting valve.

How can the rapidity of this movement be regulated? By a greater or less opening of the valve P, Figs. 143 and 144, as this valve regulates the flow of oil from one side of the dash-pot piston to the other. A slight opening causes a slow movement, while a wide opening makes possible a too rapid movement.

What would be most liable to cause breakage to the intercepting valve? Allowing the oil dash-pot to become partially or wholly empty.

What kind of oil should be used in this dash-pot? Only mineral oil, thinner, if anything, than ordinary engine oil.

What is the purpose of the key shown in the dash-pot (Fig. 144)? To prevent the intercepting valve from turning around.

With this compound, what pressure from the cab is used to operate the separate exhaust valve? Either air or steam.

Why is air pressure generally considered preferable? On account of the absence of moisture therein. As the separate exhaust valve piston *J* and its cylinder (Figs. 143 and 145) project from the front of the cylinder saddle and are exposed to currents of cold air, the use of steam therein and a lack of proper drainage might cause them to freeze in cold weather.

What objection is there to the use of air? Should the air pump stop or the pressure otherwise become exhausted, as in switching and picking up a large number of air-brake cars, there might be insufficient pressure to hold the valve open against the receiver pressure.

How is this objection overcome when air is used for this purpose? Besides the air connection to the three-way cock in the cab, there is a steam connection; closing the one and opening the other, quickly furnishes an alternative pressure for operation.

What would be the result if both the steam and the air connections were left open? There would be no effect upon the engine itself, but the steam would fill the whole air-brake system with water and seriously affect the operation of the brakes.

The Pittsburg Locomotive Works have been
one of the largest builders of two-cylinder or
cross-compound locomotives in this country.
Their location of the intercepting valve in the
saddle of the high-pressure cylinder instead of
the low-pressure cylinder is an arrangement
which obviates the necessity of a second steam
pipe for admitting live steam to the intercepting
valve and low-pressure cylinder when it is desired to work the engine non-compound.

With the valve mechanism which will be
shown and described in what follows, when the
reverse lever is either in full forward or back-up
gear and the throttle opened, the engine starts
by the admission of live steam into both cylinders and with an open exhaust passage from each
cylinder to the stack. The live steam admitted
to the low-pressure cylinder is sufficiently reduced in pressure by passing through a reducing
valve to cause the engine to have the same power
that a simple locomotive would have with two
cylinders the size of the high-pressure cylinder.
When the reverse lever is "hooked-up," or drawn
toward the center, one or more notches, it me-
chanically works an operating valve and piston
near the cab, which throws the intercepting
valve into position for compound working.
There is a hand lever in the cab, as shown in
Fig. 148, which can be used to move the inter-

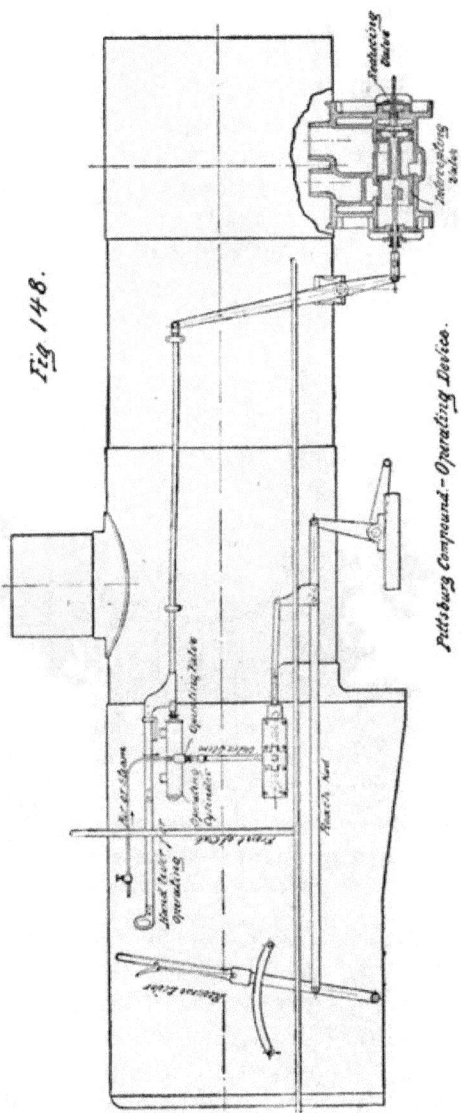

Fig. 148.

Pittsburg Compound.—Operating Device.

cepting valve in case of derangement of the operating device, or other necessity. However, the engine will only start simple when the intercepting valve is placed in proper position, as this valve does not automatically assume simple position of its own volition. Fig. 148 gives the

Fig. 149.

Pittsburg Compound.

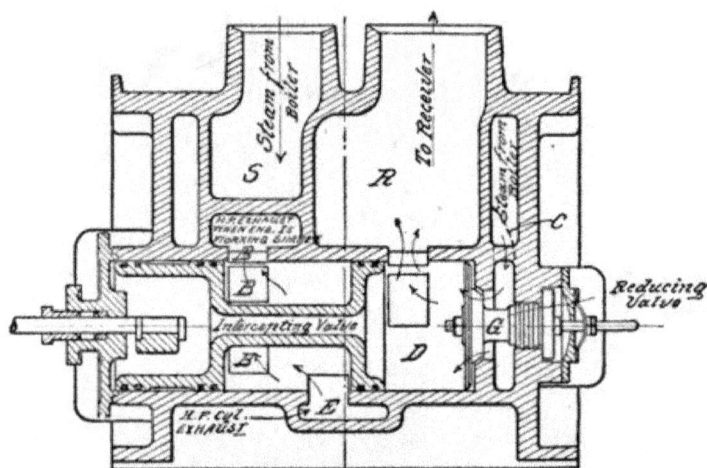

Position of Valves when Engine is working simple

general arrangement and shows the position of the intercepting and reducing valves in the saddle of the high-pressure cylinder on the right-hand side of the engine, the lever connections between the intercepting valve and the operating cylinder near the cab, and the means of moving the operating valve by its attachment to the reach-rod. The operating piston can be actuated either by air or steam pressure, the former being preferable, especially in cold climates.

Figs. 149 and 150 show on a larger scale the same section through the valves in the saddle.

Fig. 150.

Pittsburg Compound

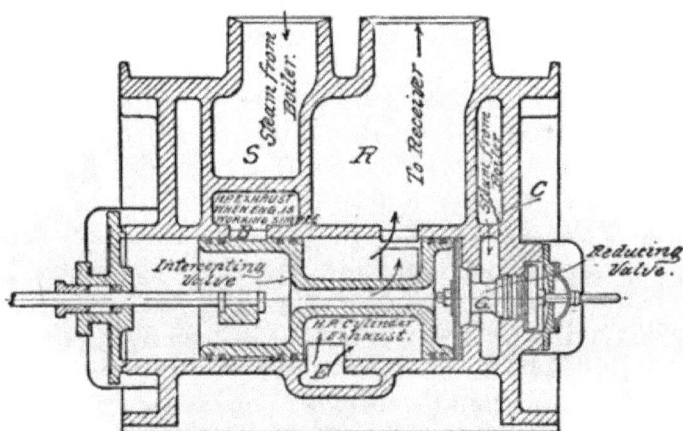

Position of Valves when Engine is working compound.

In Fig. 149 the intercepting and the reducing valves are in position for working simple, while in Fig. 150 they are shown in compound position.

Port S is the passage from the steam pipe to the high-pressure steam chest, and port E the exhaust passage from the high-pressure cylinder; port R is the opening into the receiver; port B leads to the atmosphere through the stack, being

an independent exhaust for the high-pressure cylinder.   Port C is a branch from the main steam
port S, and carries live steam to the reducing valve
G for use in the low-pressure cylinder when
needed.

When the intercepting valve is moved to the
left in position for working simple, as shown in
Fig. 149, steam from the high-pressure exhaust E
passes through the intercepting valve (between
its two piston ends) and to the atmosphere
through the independent exhaust B.   At the same
time live steam from C opens the reducing valve
G, as the valve is then unbalanced by its left end
being larger than its right, and passes through the
intercepting valve chamber D and the receiver R
to the low-pressure cylinder.   The engine is thus
made a simple engine with live steam admission
and independent exhausts for each cylinder, as
indicated by the arrows in Fig. 149.

The reducing valve G, when in use, only allows
a portion of the full pressure to pass into the
receiver, for, when the pressure in the intercepting
valve chamber D (Fig. 149) becomes equal to
about half that of the boiler, it acts against the
larger left-hand end of the reducing valve and
forces it closed—to the right—thereby throttling
the steam.

The intercepting valve moved ahead, or in compound position, as in Fig. 150, holds closed the
reducing valve G, thus shutting off the live steam
supply to the low-pressure cylinder.   At the same
time it closes the independent exhaust opening B,
instead opening the ports to the receiver R and con-

necting the high-pressure exhaust *E* therewith around the stem of the intercepting valve. As the other end of the receiver leads to the low-pressure steam chest, the high-pressure exhaust thus becomes the supply steam for the low-pressure cylinder and the engine works as a compound, the flow being indicated by the arrows, Fig. 150.

From the above description it will be observed that the office of the intercepting valve is to convert the locomotive from simple to compound, or the reverse. The duty of the reducing valve is, when the engine is working single-expansion, to reduce the live steam from the boiler to a pressure inversely proportional to the ratios of the two cylinders before delivering it to the receiver and low-pressure cylinder, thus making the crosshead loads equal on each side of the engine.

Accidents to Pittsburg Compounds.—If the engine broke down and it became necessary to disconnect either side, could the engine be run successfully on one side? Yes, it would be more powerful than a simple engine under similar circumstances and could develop good speed, inasmuch as the independent exhaust ports *B* (Fig. 149) are of considerable size.

With the high-pressure side disconnected, how should the engine be run? The intercepting valve should be placed in simple position (as in Fig. 149) by leaving the reverse lever in full gear. Live steam would then pass through the reducing valve to the low-pressure cylinder and exhaust in the usual way. The high-pressure valve should, of course, be clamped in the center.

If the low-pressure side were disconnected, how could the engine be run? Clamp the low-pressure valve in the center so as to cover all ports and work simple, with engine in full gear. Steam would then be admitted as usual through the steam pipe to the high-pressure cylinder and exhausted through the intercepting valve cavity and the independent exhaust *B*, Fig. 149.

In both of these cases what would be necessary in order to run at considerable speed? To close the small valve which supplies steam to the operating device. The hand lever can then be used for operating and thus the reverse lever hooked up to a shorter cut-off. The intercepting valve levers should be securely fastened in simple position.

With a demolished steam chest on the low-pressure side, could the engine be fixed to run in any way different from such an accident to a simple locomotive? Yes. Proceed the same as before described when the low-pressure side is disconnected, except that the reducing valve must now be held closed. This could be done by tightening up the nuts on the outside stem of the reducing valve, or by inserting washers or a block under them. This same procedure could be followed for broken valve yoke, or valve stem broken off inside of steam chest, under which circumstances it is difficult to cover ports without taking up the steam chest cover.

What should be done if the reducing valve became broken? Endeavor to start compound, that is, with the lever hooked up three or four notches, but if necessary to run simple, use very light throttle or reduce the amount of boiler pressure carried.

If the packing rings in the intercepting valve blew badly, how could it be detected? First see that the slide valves are tight, then shut off the supply of air or steam from the operating cylinder by closing the small globe valve in the cab, pull the intercepting valve clear back (by means of the hand operating lever) into simple position, as in Fig. 149, and open the throttle, still leaving the valves in the center. The blow would show in the stack. However, it is an easy matter to remove the intercepting valve, by taking off the back cap, and examine the packing rings, and this should be done frequently, that no broken pieces may catch in the ports.

With this design of compound, is it necessary to caution the engineer against the bad practice of operating the engine in simple position for too great a length of time, or at high speeds? No; for, if the small valve in the cab which supplies pressure to the operating device be open, to run the engine simple, requires that the reverse lever be at full stroke, and no competent man would work the engine thus longer than necessary.

In starting, where should the reverse lever be placed? In full motion to allow the intercepting valve to close and the reducing valve to open and admit live steam to the low-pressure cylinder.

The Richmond compound locomotive (sometimes termed the "Mellin" system, as it is built under those patents) is also of the two-cylinder or cross-compound type and belongs to the class of

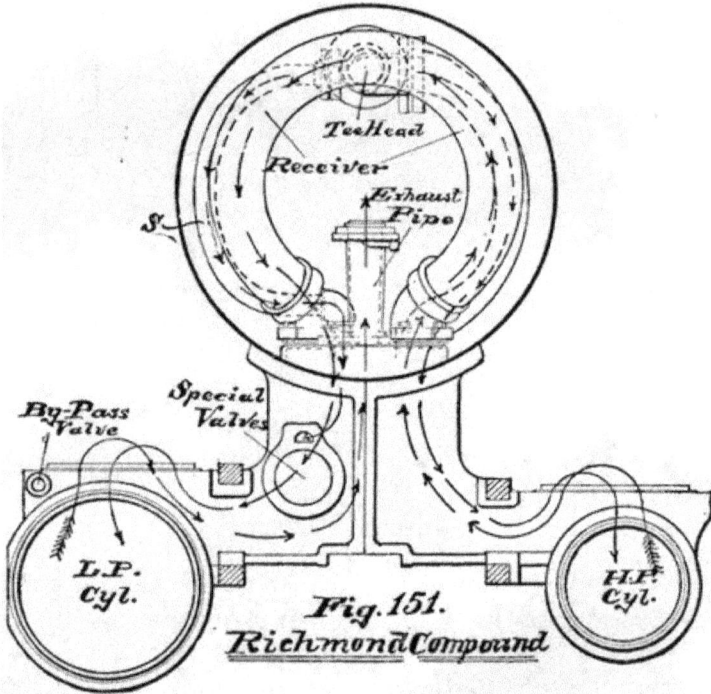

Fig. 151.
Richmond Compound

convertible compounds. The large low-pressure cylinder is placed on the right-hand side of the engine, and within its saddle, as shown in Fig. 151, is located the special valve mechanism by which the engine starts with the admission of live steam to both cylinders and thereafter automatically changes to compound, or may be converted back to simple any time at the will of the engineer. Without any movement of valves by the engineer, the locomotive is an "automatic" compound, that is, changes to working compound after the first stroke or two; but the movement by him of a three-way cock in the cab, causes the opening of a separate exhaust valve (sometimes called the "emergency valve") for the high-pressure cylinder, and thus the engine can be run simple as long as he thinks it advisable, or, if disabled, can be brought in with one side like a single-expansion locomotive.

The device called the intercepting valve really consists of three separate and distinct valves, as shown in Figs. 152 and 153.

The intercepting valve proper is marked G and is a double poppet valve with its two seats of unequal areas, and has a stem extending back and connecting with the piston P of an air dash-pot. The intercepting valve moved, or opened, as in Fig. 153, connects the receiver R with the low-pressure steam-chest port L, while if closed, as in Fig. 152, it cuts off this communication and opens the receiver R to the cavity U.

The reducing valve is a long annular valve surrounding the intercepting valve stem and closes by moving to the left. When open, it admits live steam from chamber S to the low-pressure steam chest cavity L; when closed, it cuts off this communication.

The separate exhaust or "emergency" valve E is

an ordinary bevel-seated wing-valve with its right-hand end in the form of a piston. Steam pressure from a three-way cock in the cab, if admitted against this piston, forces the separate exhaust valve open, thereby connecting the cavity U with the main exhaust cavity C; without pressure on this piston, the spring F again seats the valve.

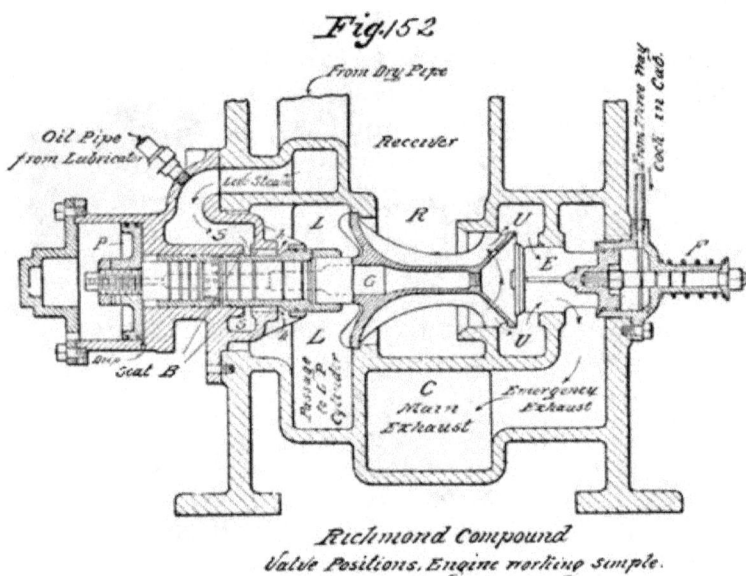

Fig. 152

Richmond Compound
Valve Positions, Engine working simple.

The reducing valve is shown heavily cross-sectioned, the intercepting valve is next as dark.

The operation of the valves is as follows: Suppose the engine to be at rest after running compound with the valves in position as shown in Fig. 153. Upon opening the engine throttle, steam passes not only to the high-pressure side in the usual manner, but also through a branch steam pipe *S* (Fig. 151) to the annular cavity *SS* (Fig. 153), bears against the shoulder *e* of the reducing valve at the seat *B* (Figs. 153 and 154) and forces the latter valve to the right, and with it the intercepting valve, to the position in which they are shown in Fig. 152. As soon as the enlarged end of the reducing valve passes the edge *h* of the

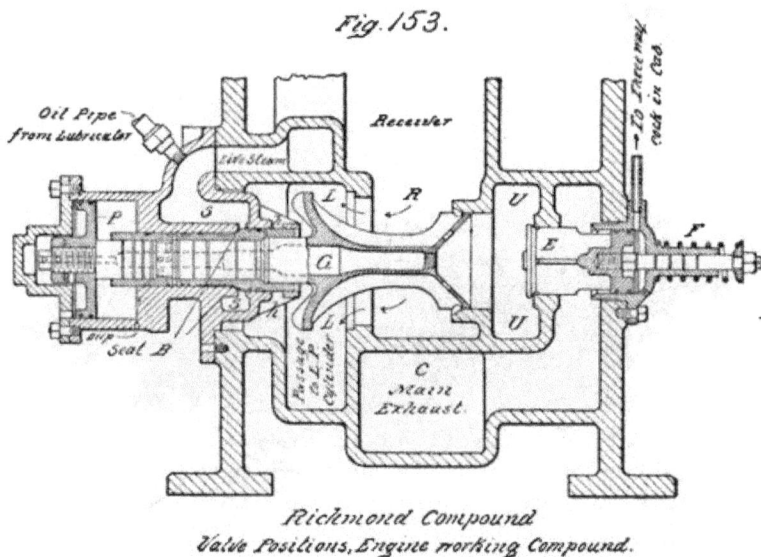

Fig. 153.

Richmond Compound
Valve Positions, Engine working Compound.

port *L* (see also Fig. 154), live steam is admitted to *L* and thence to the low-pressure cylinder, as indicated by the arrows. As we assumed the engineer had not opened the separate exhaust valve, it will remain closed instead of open, as Fig. 152 shows it; otherwise that figure would show the valve positions at this stage. Thus, while the receiver *R* is in communication with the cavity U, there is no outlet from the latter, and one or two exhausts from the high-pressure cylinder into the closed receiver R will produce sufficient pressure to act against the larger left-hand face of the intercepting valve G and cause it to move to the left, shoving with it and closing the reducing valve, as shown in Fig. 153. Now the receiver pressure becomes the supply for the low-pressure cylinder as shown by the arrows, and the engine works compound thereafter. The engine will start all ordinary trains in this manner but if the train

Fig. 154

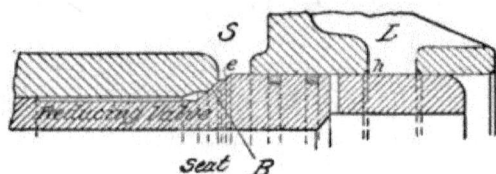

to be started is a very heavy one, or such a train is threatened with stalling while ascending a heavy grade, the handle of the three-way cock in the cab should be turned to open the separate exhaust valve E. This would vent the pressure in chamber U to the main exhaust C, as indicated by the arrows in Fig. 152. If the engine were starting, no pressure could accumulate in the receiver R on account of this vent to the atmosphere, and hence all valves would remain as in Fig. 152 and the engine would continue to work simple until such time as the engineer closed the separate exhaust valve E by means of his cab valve. If, however, the engine were stalling working compound (Fig. 153) and the engineer opened the

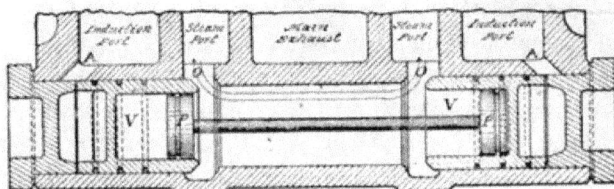

Fig. 155.
The Richmond Compound
By Pass Valve

Position when Running
Steam shut off

separate exhaust valve E, the removal of pressure from chamber U would cause the intercepting valve to close, assisted by the live steam pressure on the reducing valve shoulder, and thus all valves

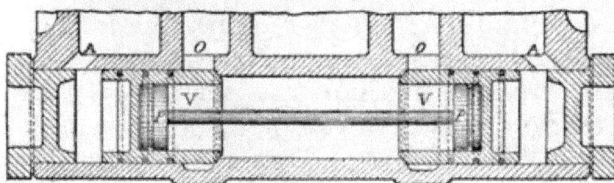

Fig. 156.
The Richmond Compound
By Pass Valve

Position when running
Under Steam

would remain as in Fig. 152 until the engineer desired to work compound again. Then he would simply close the separate exhaust valve E

and permit the valves to automatically assume compound working by the accumulation of pressure in the receiver *R*.

The annular shaped reducing valve appears so much like an inoperative sleeve around the intercepting valve stem that the separate sketch is made on a larger scale in Fig. 154, in order to illustrate its principle the more clearly. Live steam from *S* bearing against the shoulder *e* would force the reducing valve from the position in the cut to the right until shoulder *e* had passed the edge *h* of the chamber *L* and live steam thus be admitted to *L* and the low-pressure cylinder. As soon as the pressure in *L* became sufficiently great, say about half that in *S*, the valve would close partly so that only the desired proportion of the full boiler pressure would be admitted to the low-pressure cylinder.

The injurious action of large pistons in pumping air when the engines are shut off, as previously mentioned under "Classes of Compound Locomotives and their General Construction," is prevented in compound engines of the Richmond type by the use of "by-pass" valves located in the casting of low-pressure cylinder, as shown in Fig. 151, and further shown in detail in their two positions by Figs. 155 and 156. They work automatically by the opening and closing of the engine throttle, their construction and operation being as follows: The outer ends of the two piston valves *VV* are connected by the passages *AA* with the induction ports of the low-pressure cylinder and their inner ends are connected by the passages *OO* with the

steam ports. The pistons *PP* simply act in the
cavities of the valves as dash-pots to prevent slamming of the valves.

Open the engine throttle, and live steam from
the induction ports through the passages *AA*
drives the valves *VV* toward each other against
their seats and communication between ·the
two steam ports *O* and *O* is closed, as in Fig. 156.
With the throttle closed, the least vacuum in the

Fig. 157.

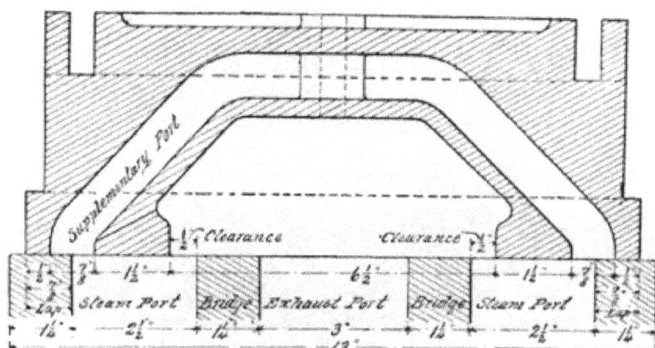

Brooks Patent Slide Valve, In Central Position

steam chest. acting at *AA*, is aided by compression
in the cylinder through ports *OO*, and the "by-
pass" valves move outward to the position shown
in Fig. 155. In this position the two steam ports
OO are connected and air passes freely from one
side of the piston to the other—the purpose for
which they are applied.

The necessity for very large port openings for
the large low-pressure cylinders was mentioned
under the general discussion of compound loco-

motives and it was there stated that auxiliary
ported valves of the "Allen" type were used to
give a double port opening to the cylinders in the
early part of the admission period.

One Richmond design of slide valve employs
this auxiliary port not only for double early admission, but also to give a double opening in the
early exhaust period.   Fig. 157 shows the design
and gives the dimensions.   It will be noticed that

Fig. 158.

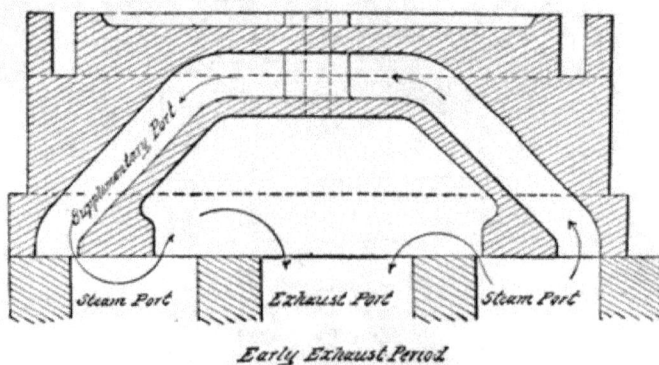

Steam Port   Exhaust Port   Steam Port

Early Exhaust Period

the "Allen," or supplementary, port is much wider
at the seat than usual.*

To clearly show the features of novelty in this
valve over the ordinary slide valve, Figs. 158 and
159 are given.   The arrows in Fig. 158 indicate the
double exhaust from the right-hand end of the
cylinder during the early exhaust period.   In Fig.
159 the arrows indicate similarly the ordinary

*Letters patent have been granted for this valve.

action of an "Allen" ported valve in giving double
admission to the same end of the cylinder during
the early steam period. This style of slide valve
has been in use on locomotives and given only a
six-inch extreme travel of valve and has proven
very successful in keeping down the back pressure, especially at high speeds.

Fig. 159

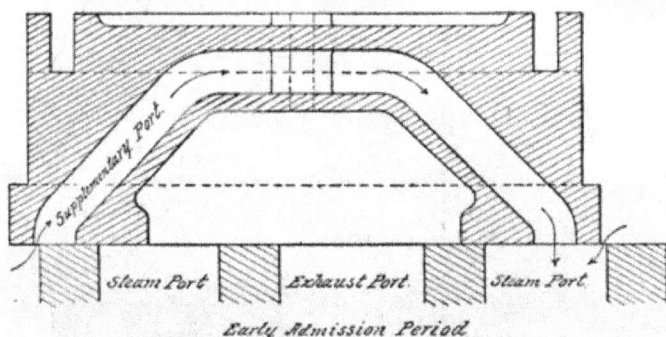

Steam Port   Exhaust Port   Steam Port

Early Admission Period

Accidents to Richmond Compounds.—In case of
accident on the road requiring the engine to be
run in with one cylinder, what should be done?
Ordinarily, open the separate exhaust or "emergency" valve and do nothing otherwise differently
than with a simple engine.

In case the engine was running with the low-
pressure side only, why do you instruct to open
the separate exhaust valve? So that there shall
be no accumulation of pressure in the receiver
which might occur in case of a leaky slide valve
or balance strips leaking on the high-pressure
side.

As the emergency port openings are small, what could be done in order to obtain considerable speed in case the break-down occurred in passenger service? Try to block the slide valve on the disabled side in an extreme position so as to uncover the exhaust port.

What should be done in case the reducing valve stuck open or became broken? Care should be used in starting by opening the throttle very

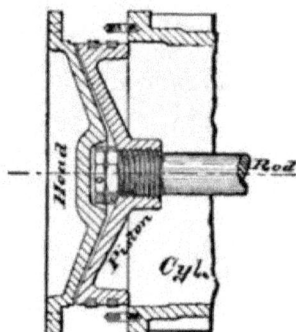

Fig. 160.
-Piston for Lightness & Strength-

slightly and the engine should be run as a compound only. If it had to be run simple, the boiler pressure should be reduced about one-half.

What might cause the reducing valve to stick? Feeding valve oil to the low-pressure side when the engine is working compound and there is consequently no flow of steam through this valve.

What precaution would it be advisable to observe in case the intercepting valve became broken? The engine should be run compound only and it would be safer not to stop with the low-pressure side on center. If this latter precaution were not observed, the high-pressure piston might have considerable back pressure upon it from the receiver, should the intercepting valve break in such a way as to permit live steam from the reducing valve to enter the receiver in starting.

What prevents slamming of the intercepting valve? The air dash-pot connected thereto.

In this class of compounds, why are the cylinder heads sometimes dished, or not flat, as is usually the case? In order to give a maximum strength with a minimum weight of piston, these builders frequently use a piston such as is shown in Fig. 160; hence the cylinder heads conform to the shape of the piston.

The intercepting valve originally employed by the Rogers Locomotive Works for their system of two-cylinder compound, was placed in the smoke-box and was controlled automatically and beyond the will of the engineer. In starting, by placing the reverse lever in either extreme position, a separate reach-rod with suitable levers was made to open the reducing valve and admit live steam from the boiler to the low-pressure cylinder, and close communication between the latter and the receiver, so that the high-pressure cylinder exhausted into a closed reservoir or receiver. After one or two strokes sufficient pressure accumulated in the receiver to automatically throw the intercepting valve into position for working compound and the hooking up of the reverse lever closed the reducing valve.

The succeeding arrangement used by these builders is fully described and illustrated in what follows. The device is for a two-cylinder or cross-compound locomotive, and is placed in the high-pressure saddle on the right side and connected to a lever in the cab having three notches, and is operated by the engineer at will. As long as this lever remains in the forward notch the engine works as a single expansion locomotive; the back notch places the intercepting valve in compound position, while the middle notch makes an automatic compound, that is, the intercepting valve works the engine simple for a few strokes and then the valves automatically assume the compound position. Thus this engine belongs to the class of convertible compounds.

The component parts of the device are: A regulating valve (positively controlled by the cab

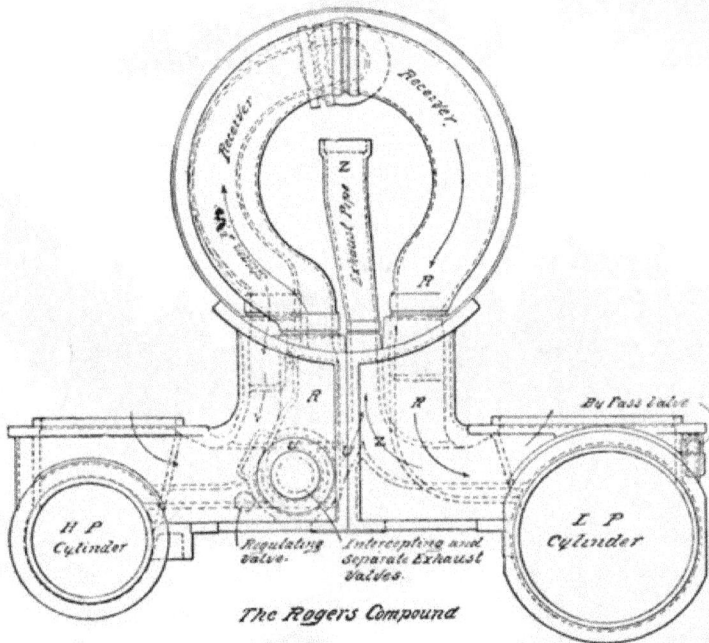

Fig. 161

The Rogers Compound

lever above mentioned), an intercepting and a reducing valve (controlled either positively by the regulating valve or at times automatically by the pressure in the receiver), and a separate exhaust valve (controlled by the operating valve). The

Fig. 162.

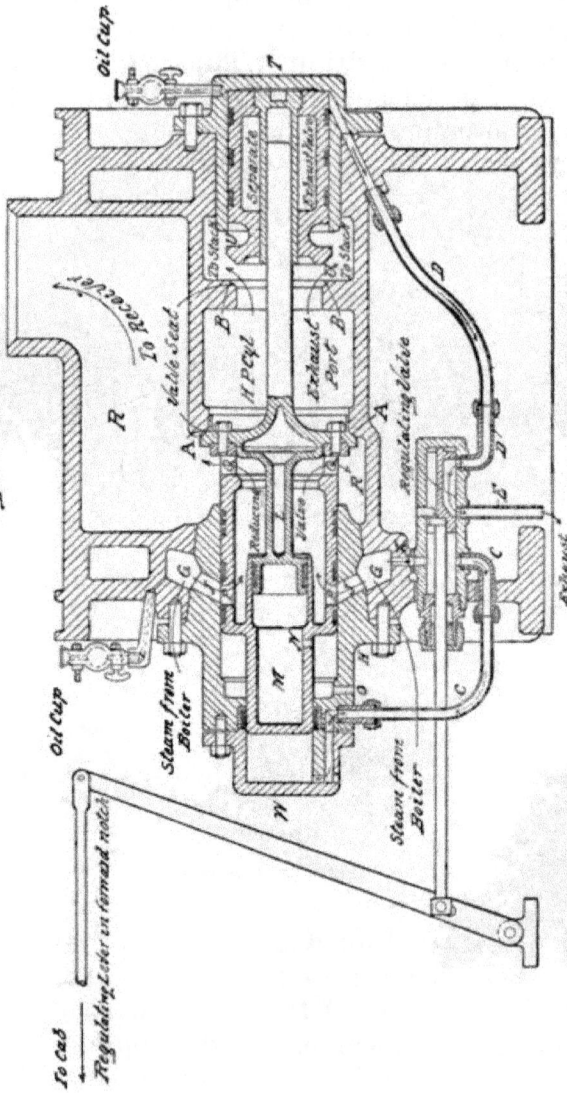

Rogers Compound
Position of Intercepting valve when running simple

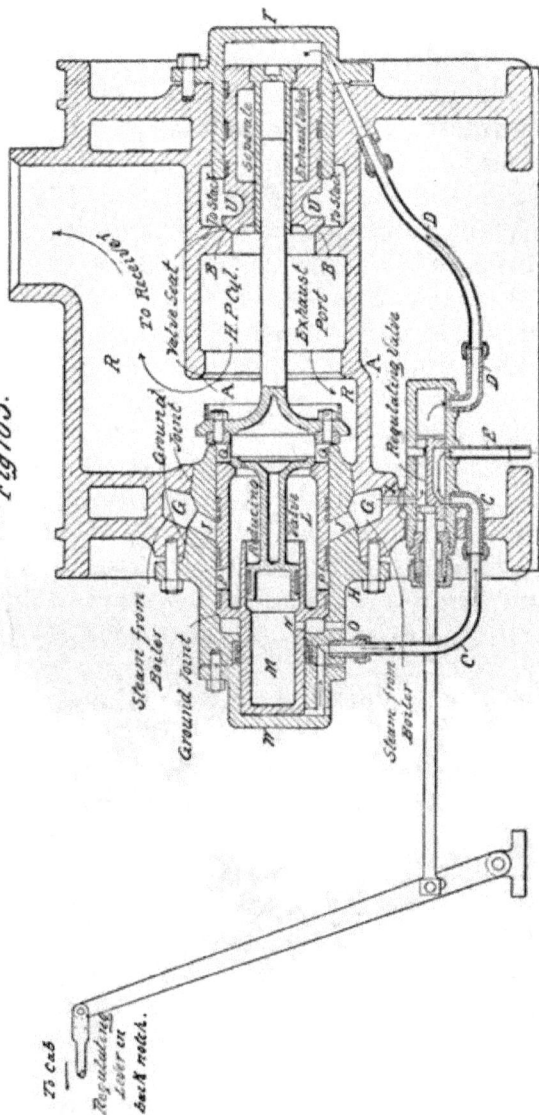

Fig. 163.

Rogers Compound

Position of Intercepting Valve when running compound

intercepting valve has two positions, one for working compound and one for working simple. In
the former position, as shown in Fig. 163, the high-
pressure cylinder exhaust is connected to the receiver and thence to the low-pressure side, while
in the simple position, as shown in Fig. 162, this
communication is closed and the high-pressure
exhaust directed through the separate exhaust
valve to the stack, live steam at the same time
being admitted to the low-pressure steam chest
through the reducing and the intercepting valves.

To somewhat reduce the high compression that
takes place in the large low-pressure cylinder
when running at high speeds without steam—as,
for instance, down grades—a by-pass arrangement
similar in some respects to "La Chatalier" brake is
employed. This consists of an automatic device
for connecting the two sides of the low-pressure
piston when steam is shut off, and will be fully
described hereafter.

Fig. 161 is a view from the front, showing the
high-pressure cylinder to be on the right-hand
side of the engine. In the saddle on that side
are shown the positions of the regulating valve
and the intercepting and separate exhaust valves.
The location of the by-pass valve on the low-
pressure cylinder is also shown.

Figs. 162 and 163 are sectional views of the
special valve mechanism. For simplicity in ex-
plaining their working, the regulating valve is
shown as though it were located directly underneath the intercepting valve which is not its actual
position shown in Fig. 161. The intercepting

valve is composed of two parts bolted together and is shown cross-sectioned more heavily than its casing while the reducing valve *L* within it is distinguishable by its still darker appearance. Its principal port openings to the receiver and from the high-pressure cylinder exhaust are lettered and the direction of flow through them is indicated by arrows, that an understanding may be the more easily arrived at. Port *G* is an annular chamber around the intercepting valve cylinder *H* and is connected with the high-pressure steam pipe in the saddle and therefore has live steam pressure when the engine throttle is open. Between *G* and the intercepting valve is a series of holes marked *J*. The intercepting valve is hollow and has two series of holes through its shell at *P* and *Q*, the latter being to the right of the reducing valve seat.

Fig. 162 shows the intercepting valve closed (to the right) on its seat *A*, and the reducing valve off its seat (open), as it is when working simple except that the reducing valve *L* is then brought near enough to its seat to the left, to prevent more than about one-half the boiler pressure from passing through it. As usual, the reducing is accomplished by the two ends of the reducing valve *L* being of different diameters, the smaller end having its outer left-hand side always connected to the atmosphere through small ports marked *N* and *O*, while the outside of the larger right-hand end is exposed to the receiver pressure when in operation, as in Fig. 162. The left-hand smaller end *M* of the intercepting valve works in a cylinder *W*, while the right-hand

larger area of the intercepting valve is exposed
either to the receiver or the high-pressure exhaust
cavity, according to the position of the valve.
The separate exhaust valve is located in a chamber *U* directly connected to the main exhaust
pipe *Z*. (See Figs. 161 and 163.) The right-
hand end of the separate exhaust valve is in the
form of a piston and works in a cylinder *T*. The
outer ends of these two cylinders, *W* and *T*, are
connected by small pipes *C* and *D* with the regulating valve, as shown in Figs. 162 and 163. The
stem of the intercepting valve extends over into
separate exhaust valve and forms a guide to
keep both valves central, and also serves as a
dash-pot, but allows each valve to act independently of the other. All pistons are fitted with
packing rings, as shown, and the valve seats and
ground joints are so indicated.

The regulating valve receives steam from *G*
through a small port *X* and therefore has live
steam whenever the engine throttle is open. The
regulating valve consists of a slide valve of the
well-known "*D*" type and moves on a seat having
two steam ports and one exhaust opening, as
usual with a slide valve. Steam admitted through
the right steam port and pipes *DD* to chamber *T*
would bear against the separate exhaust valve and
close it, as in Fig. 163. Steam through the left
port and pipes *CC* to chamber *W* would force the
intercepting closed, as in Fig. 162. If the engineer places the regulating valve lever in the
forward notch, as in Fig. 162, live steam entering
the regulating valve chamber from ports *G* and *X*

finds the left hand port uncovered and passes through the small pipe *CC* to the cylinder *W* (as indicated by arrows), moving the intercepting valve to the right against its seat *A* and thereby closing communication between the high-pressure cylinder exhaust port and the receiver *R*. In this position live steam pressure from *G* passes through ports *JJ* and *PP* to the interior of the intercepting valve, automatically opening the reducing valve and flowing through ports *QQ*, enters the receiver *R*, the other end of which is connected to the low-pressure steam chest, as shown in Fig. 161. Thus the low-pressure cylinder has received live steam at a reduced pressure and can work independently of the high-pressure cylinder, as its exhaust is as always to the stack. The intercepting valve now being closed, the exhaust from the high-pressure cylinder (its communication to the receiver being thereby shut off) forces the separate exhaust valve off its seat, as in Fig. 162, and escapes to the stack and the engine works simple, or with a live steam admission and an exhaust to the atmosphere on both sides. The separate exhaust valve remains open, as shown in Fig. 162, because chamber *T* is connected to the atmosphere through pipe *DD*, the under side of the regulating slide valve, and pipe *E*—as indicated by arrows.

If the regulating lever in the cab be moved to the back notch, the regulating valve takes the position shown in Fig. 163 and will then admit steam through the right-hand pipe *DD*, into

chamber *T*, pushing and holding the separate exhaust valve against its seat *B*, and at the same
time chamber *W* will be in exhaust through pipes
CC and *E*, as indicated by arrows. The separate
exhaust valve being now closed, a slight pressure
in the cavity marked "high-pressure cylinder
exhaust" will open the intercepting valve, that is,
move it off its seat *A* into the position shown in
Fig. 163. Then the high-pressure cylinder exhausts through the intercepting valve into the
receiver and thence to the low-pressure steam

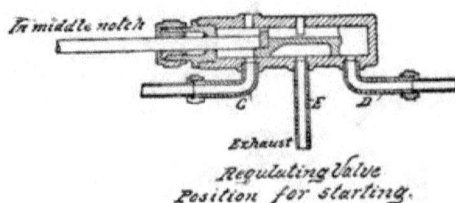

Fig. 164

Rogers Compound

In middle notch

C   E   D

Exhaust

*Regulating Valve
Position for starting.*

chest, and the engine works compound. No live
steam from ports *GG* can get to the receiver in
this position as the ports *PP* are not in register
with those at *JJ*.

It will be noticed that there is an extended flange
on the face of the intercepting valve, that enters the
seat of the latter enough before closing and leaves
it sufficiently late in opening, to prevent live
steam from flowing through ports *PP*, *QQ*, and *R*
to the high-pressure cylinder exhaust port when

the intercepting valve is opening or closing. If the engine is standing and the regulating valve is moved to its central position, as shown in Fig. 164, when the throttle is opened steam will pass through both pipes *CC* and *DD* to the outer ends of the intercepting and the separate exhaust valves respectively, and they will both be instantly closed against their respective seats *A* and *B*, Figs. 162 and 163. The receiver would then be filled with steam through the hollow intercepting valve and the reducing valve, as indicated by the arrows in Fig. 162; but as soon as the pressure at the high-pressure exhaust port became about twenty per cent. of the pressure in the dry pipe at the time, the intercepting valve would be opened and thereafter held open by the receiver pressure acting toward the left on the large right-hand end of the intercepting valve; thus a slight pressure on the large area of this valve overbalances the higher pressure on the smaller left-hand end *M* in chamber *W*, and the engine automatically changes to compound after two or three strokes, the main ports being opened as in Fig. 163.

If the engine had been working simple, with the valves as in Fig. 162, and the operating valve lever were placed in either the center or back notch, the working would automatically be changed to compound; if in the center notch, it would take a slight accumulation of pressure at the high-pressure exhaust port, as just explained; if in the back notch the intercepting valve opens with almost no pressure from the high-pressure exhaust port as then chamber *W* is in exhaust through pipe *CC*

and the regulating valve.   Subsequently to working simple, it is best to move the regulating lever
to its middle position (Fig. 164) for a few sec-
onds without closing the engine throttle; after
that it can be moved to its back notch, as in Fig.
163, and left there until it is desired to again
start simple or go from compound to single-ex-

Fig 165.

Rogers Compound

Fig. 166.

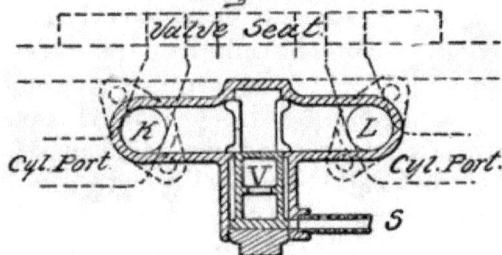

pansion in order to prevent stalling on a heavy
grade.
    Figs. 165 and 166 show two views of the by-
pass arrangement which is placed on the low-pressure cylinder in the position shown in Fig. 161.
It consists of a valve V which automatically opens

and closes the port in a hollow casting bolted to the cylinder and connected below the valve seat at K and L with the steam ports of the low-pressure cylinder, as illustrated in Figs. 165 and 166. The hole through this by-pass casting is two inches in diameter. In Fig. 166 the valve V is shown dropped down in its chamber, thus leaving a two-inch open communication between the steam ports and hence the two sides of the low-pressure piston. A small pipe from the live steam port in the saddle of the high-pressure cylinder leads through the pipe S to the under side of this cylindrical valve V, as shown, so that when the engine throttle is opened the valve V raises to the top of its chamber and shuts off all communication between the ports K and L, remaining there as long as the high-pressure cylinder is receiving steam; but when the throttle is closed, it falls again by gravity and is thus an automatic valve.

Accidents to Rogers' Compounds.—In case of break-down on one side of this engine, can it be run in with the other side? Yes. The reducing valve would furnish a live steam supply for the low-pressure cylinder, or the separate exhaust valve would give an exhaust opening for the use of the high-pressure cylinder, according to which side was to be used; hence, in either case, general instructions would be to place the regulating valve in its extreme forward position, as for working simple, Fig. 162.

Under such circumstances would the engine be as powerful as a simple engine on one side? Yes, considerably more so at low speed, but the smaller port openings than usual would not permit of as great speed.

As there are three notches in the cab for the regulating valve lever, how should this be handled when it is desired to change from simple to compound, or the reverse? It is not advisable to throw it from forward to back notch, or the opposite, without pausing a second or two in the middle notch and thereby producing a more gradual movement and causing a cushioning of the intercepting and separate exhaust valves.

Under ordinary circumstances, in what position is it best to keep the regulating valve? In the back notch, or compound position.

How should a light train be started? With the regulating valve lever in the center notch. The engine starts thus as an automatic compound.

If the engine actually works compound in the center position, why is that not advisable? Because in that position the opening and closing of the throttle while running and the wide variations of pressure in the steam pipe and receiver when running, would cause more or less unnecessary movements of the intercepting valve.

When running the engine with one side, in what position will the "by-pass" valves (Figs. 165 and 166) be found? No matter which side is in use, live steam from the high-pressure steam pipe acts to keep them closed.

What kind of oil should be used in the oil cups shown in Fig. 162? Nothing but valve oil and that regularly but not too often, as it has a tendency to gum the valves where there is not a constant flow of steam.

If the high-pressure side were on the center, why would not the engine move with the operating lever in compound position (back notch)? Because then the valves would be as in Fig. 163 and no live steam could be admitted to the low-

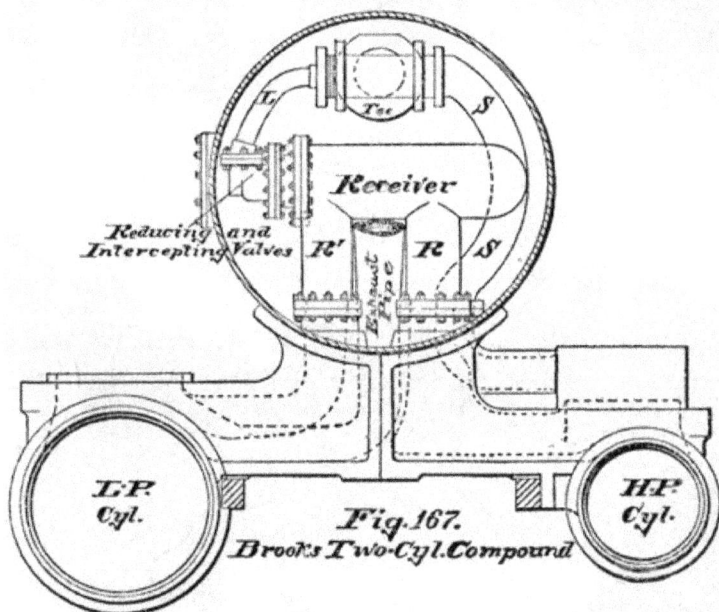

Fig. 167.
Brooks Two-Cyl. Compound.

pressure cylinder. This would not be the case in starting with the regulating lever in either of the other two notches.

BROOKS TWO-CYLINDER COMPOUND.

This design is of the two-cylinder or cross-compound type, built under the Player patents, having the high-pressure cylinder on the left side and a receiver in the smoke-box between the two cylinders, as shown in Fig. 167.

There is a combined admission, pressure-regulating and intercepting valve located either on the receiver in the smoke-box as shown, or in the cylinder saddle, which valve upon opening the engine throttle admits live steam at a reduced pressure to the low-pressure cylinder; at the same time the intercepting valve automatically closes, preventing the live-steam pressure from working against the high-pressure piston. The reducing valve remains open until the pressure in the receiver, accumulating from the high-pressure exhaust, becomes equal to or slightly greater than that on the low-pressure side, when the valve automatically closes, thereby shutting off the admission of live steam, and the intercepting valve simultaneously opens the receiver to the low-pressure steam chest, and the engine works compound thereafter as long as the throttle remains open.

In order to give the engineer control of the locomotive at all times, controlling valves are provided on the low pressure side. In the illustration (Fig. 168) these valves, *A* and *B*, are shown located in the bottom of the receiver and are connected to the cab by suitable levers *C* and *D*. They are sometimes made larger, connected higher up with the exhaust pipe, and arranged as to work automatically in combination with the intercepting valve, so that the engine can be run simple as long as desired and the exhaust take

Fig 168.

Test head

S

L

Receiver

H. P. Steam Pipe

Exhaust Pipe

Reducing and Intercepting Valves

To Cab

Receiver

To Cab

D

C

D

A

B

C

L. P. Cylinder

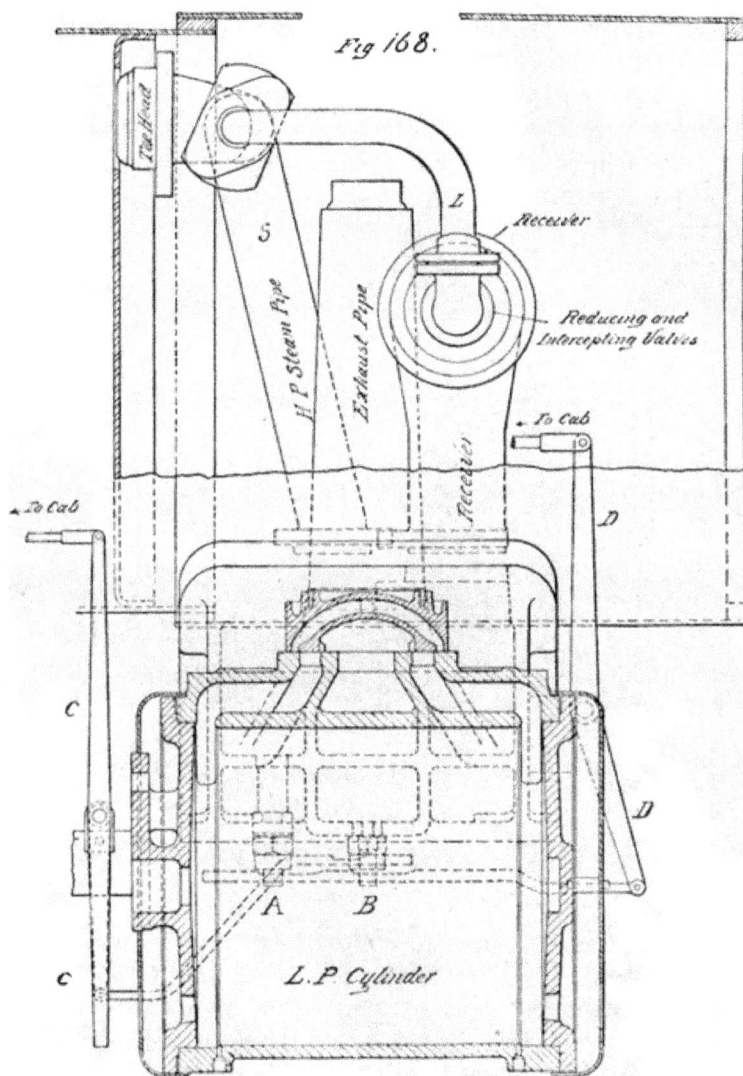

BROOKS TWO-CYLINDER COMPOUND.

place through the stack; however, its builders
have claimed the design here shown to be "all that
is necessary to give the engine its maximum
power."

The operation of the locomotive is as follows:
When the engine throttle is opened, live steam is

Fig. 169.

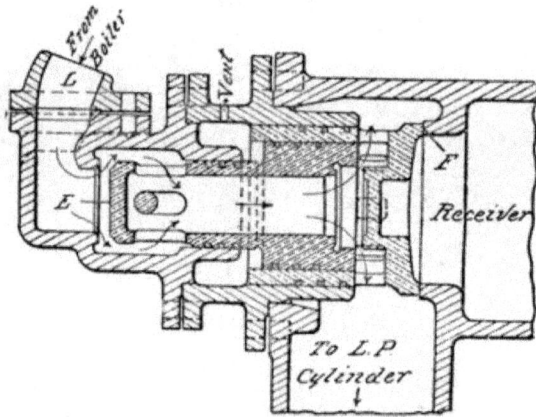

Brooks Two-Cylinder Compound
Starting Position
Reducing Valve Open, Intercepting Valve Closed.

admitted to the high-pressure steam chest through
the steam pipe S (Figs. 167 and 168) and operates upon the high-pressure piston in the usual
manner. At the same time, through the small
steam pipe L, steam acts against the seat E of
the reducing valve, then closed as in Fig. 170,

causing this valve to open, passing through the
hollow portion of the valve, automatically closing the intercepting valve against its seat *F* on
the receiver, as in Fig. 169. Steam then flows
through the passages in the intercepting valve
and down to the low-pressure steam chest, as in-

Fig. 170.

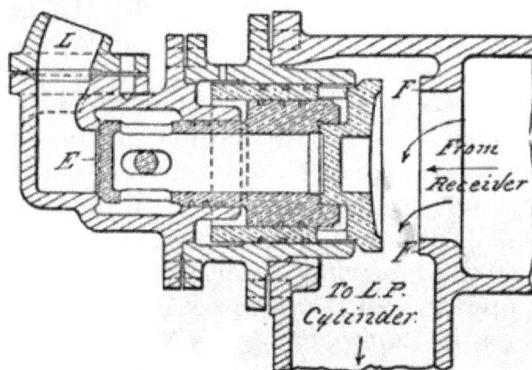

Brooks Two-Cylinder Compound
Compound Position
Reducing Valve Closed, Intercepting Valve open.

dicated by the arrows. To render them readily
distinguishable, the reducing or regulating valve
is cross-sectioned very heavily, while the intercepting valve is less dark in appearance.

From Figs. 169 and 170, it will be seen that
the right-hand end of the reducing valve is of
larger area than that acted upon by live steam at

its seat on the left-hand end *E*.   It is this difference that causes the partial closing of the
valve, thereby throttling the steam passing
through it to a reduced pressure.   These areas
being about two to one, the reducing valve, in
order to equalize the work of the two cylinders
in starting, prevents more than about one-half
the pressure in the dry pipe from passing to the
low-pressure cylinder.

As soon as the high-pressure cylinder has exhausted sufficient steam into the receiver to
overbalance the reduced live steam pressure
holding the intercepting valve closed, this valve
opens automatically, at the same time locking the
regulating valve against its seat as shown in Fig.
170.   Exhaust steam from the high-pressure cylinder then flows through the receiver to the low-
pressure steam chest, as indicated by the arrows
and the engine works compound.   The receiver
pressure can become considerably reduced and
still, through the action of the combined valves,
keep the pressure regulating valve closed.

Accidents to Brooks Two-Cylinder Compounds.—
What should be done in case of an accident
necessitating the removal of the main rod on the
low-pressure side?   Disconnect that side, block
the crosshead and clamp the valve in center, as
with a simple engine.   Open wide the controlling valves underneath the saddle and run in with
one side.   The exhaust will take place through
these valves.   To obtain a larger exhaust opening
than the controlling valves furnish and thus permit of more speed and better steaming qualities,

attempt should be made to place the slide valve
on the disabled side so as to uncover the exhaust
port and thus give an exhaust through the stack.

What should be done for a similar accident
to the high-pressure side? Just the same as for
a simple engine. There being then no exhaust
into the receiver, the intercepting and reducing
valve would admit live steam to the low-pressure cylinder upon opening the engine throttle.

What mode of procedure should be followed in
case of a broken valve stem or rocker arm? Proceed as above, according to which side was broken.

What should be done if the intercepting
valve became broken so as to leave an opening between both ends of the receiver? The
proper method would be to disconnect the small
steam pipe *L* (Figs. 167 and 168) and insert a
blind gasket. The engine could then be started
and worked only as a "strictly plain" compound
and should not be stopped with the high-pressure
side on center, as no live steam could be given
the low-pressure cylinder for starting. Without
disconnecting the steam pipe the engine could be
run in with part of a train by reducing the boiler
pressure or by using a slightly open throttle, in
which case live steam from the reducing valve
would pass through the receiver and work against
the high-pressure piston in starting, and hence
the engine should not then be stopped with the
low-pressure side on center.

What should be done for a broken reducing
valve? The same as for a broken intercepting
valve.

Should the controlling valves fail to open, where would you look for the cause of the trouble? They operate very similarly to cylinder cocks and their levers becoming disconnected or bent may cause them to either fail to open, or remain continually open, as the case may be. The remedy is obvious.

BROOKS FOUR-CYLINDER TANDEM COMPOUND.

The Brooks (Player System) of four-cylinder compound is of the tandem type, that is, two cylinders on each side, one ahead of the other, as shown in Fig. 109. With this system the two low-pressure cylinders and their saddles are placed similarly to ordinary single-expansion cylinders, with which they can always be made interchangeable. They are of course larger, have a different style of steam chest and, bolted preferably to their forward ends, as shown in Figs. 109 and 171, are the two high-pressure cylinders which have steam chests communicating with the steam chests of the low-pressure cylinders with an enlargement between, all together forming a receiver. There are no devices in the smoke-box except those usual to single-expansion locomotives.

The steam is supplied to the high-pressure valve chest through suitable pipes connecting with the usual steam pipes in the smoke-box, and the exhaust from the low-pressure cylinder is through the usual cavity in the saddle.

As shown in Fig. 171, the low-pressure cylinders are fitted with the usual balanced slide

valves, while the valves for the high-pressure cylinders are of the piston type and hollow, having
internal admission and external exhaust edges.
The low-pressure valve is controlled by the usual
eccentric and rocker arms, as ordinary with simple
engines. It is rendered advisable to have an internal admission with the high-pressure valve in
order to lessen the cooling of the steam from the
boiler as well as for constructive reasons.

On account of the high-pressure valve having
internal admission and the low-pressure valve
having external admission (the latter being
usual with ordinary slide valves) it is necessary
that these valves should travel in opposite direc-
tions. This is accomplished by placing in the
receiver (as shown in Fig. 171) an intermediate
rocker arm A which is connected by rods to both
valves and has arms of the desired ratio to give
a relatively less travel to the high than to the
low-pressure valve. These valves are so propor-
tioned that when running in nearly full gear, the
high-pressure cut-off takes place later in the
stroke than the low-pressure cut-off, but when the
engine is hooked up the relative time of cut-offs
changes.

The pistons of both the high and the low-
pressure cylinder are fitted upon the same piston
rod, the intermediate head between the two
cylinders having suitable metallic packing, as
shown in the illustration.

On the low-pressure steam chest is fitted a reducing and starting valve which is connected with
the high-pressure steam pipe, as shown with the

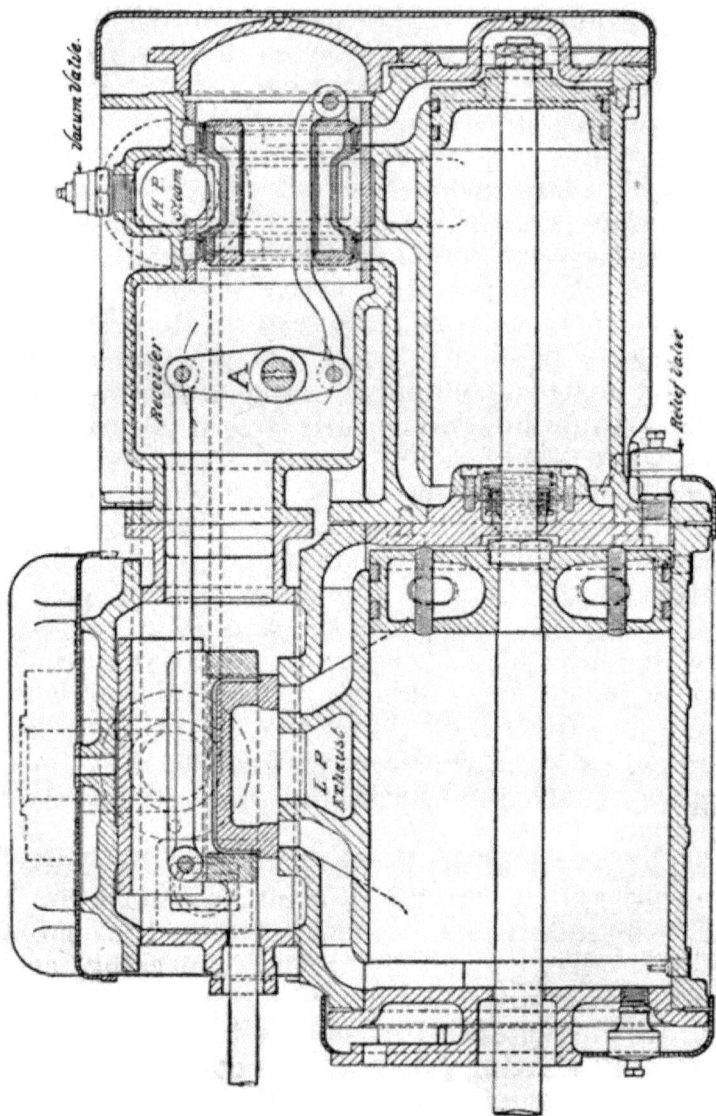

Fig 171.

Brooks Tandem Four-Cylinder Compound.

reducing valve in closed position by Fig. 172.
By the connection of a rod with the arm of the
lift shaft, this reducing valve is automatically
opened whenever the reverse lever is placed
either in full forward or full back gear. In the
intermediate positions of the reverse lever, this
reducing valve is locked to its seat by a suitable
spring, so that it is rendered inoperative and the
engine must necessarily work compound at all

Fig. 172

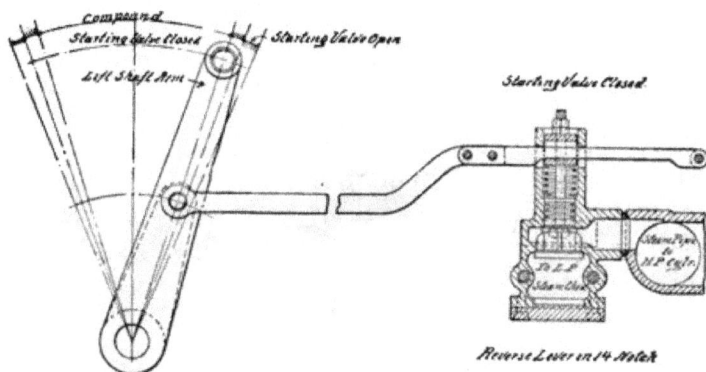

Brooks Tandem Compound
Operation of Starting Valve

times and under all conditions of steam pressure
when the reverse lever is in any other position
than full gear. Fig. 172 shows the range of the
lift shaft arm for the starting valve to be either
open or closed. This combined starting and reducing valve, when open, permits live steam to
enter the low-pressure cylinder at a pressure
equivalent to the maximum pressure obtained in

this cylinder when the engine is working compound.   As soon as the engine has made one
complete revolution, the receiver becomes charged
sufficiently by the exhaust from the high-pressure
cylinder to close the reducing valve against its
spring, thus automatically rendering the starting

Fig. 173.

Brook's Tandem Compound.

Starting Valve Open.

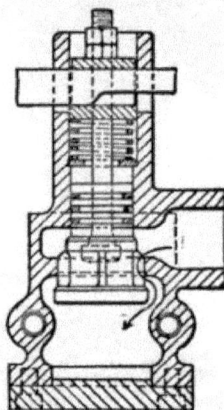

Reverse Lever in full Gear, 17 Notch.

valve inoperative and thereby necessitates the
compound working of the engine.   The reducing
valve on a larger scale is shown in open or starting position by Fig. 173.

The engine operates in the following manner:
Steam is admitted to the high-pressure steam

chest through suitable pipes, into the annular steam admission cavity surrounding the high-pressure piston valve, thence to the high-pressure cylinder, and exhausts into the receiver—the exhaust from the forward end of the cylinder passing through the inside of the hollow piston valve. The low-pressure steam chest, being also of large size, somewhat increases the receiver capacity, so that practically uniform pressure is maintained therein, steam from the receiver being admitted to and exhausted from the low-pressure cylinder by an ordinary slide valve which gives a distribution in that cylinder in the same manner as for simple engines.

This style of compound has been in operation for several years and, the builders state, shows an excellent economy in fuel, water, and repairs.

Accidents to Brooks Four-Cylinder Tandem Compound.—What would you do if it became necessary to disconnect one side of the engine? The same as with a simple engine of similar design in other respects.

In case the front head of the high-pressure cylinder became broken, how would you proceed in order to run the engine in with a full train? Remove the front steam chest head, place the reverse lever so as to throw the high-pressure valve clear ahead, disconnect the high-pressure valve and its intermediate rocker-arm rod (Fig. 171) and block this valve securely in the center so as to cover all ports; then block the starting valve open on that side, thereby allowing live steam at a reduced pressure to enter the low-pressure cylinder. This will enable the engineer to slowly handle a full train, but should not be so run for any distance or the high-pressure cylinder will be badly cut. In the lighter service of passenger trains, the engine would be capable of greater speed by disconnecting the broken side, blocking the valves in the center of their seats, and running with one side, as would be done with a simple engine.

If the intermediate rocker arm or the valve rod should break what should be done? Broken at some certain points, the engine might be treated as for broken high-pressure cylinder head; but the preferable procedure, all things considered, would be to disconnect on that side and be sure to block both valves over ports, to do which it may be necessary to remove front steam chest head and block the high-pressure valve inside.

In disconnecting one side of these engines, should it be imperative to securely block the crosshead? It should, because of the difficulty of securely holding the high-pressure valve in its central position.

How could it be determined if the metallic piston-rod packing between the two cylinders were blowing badly? Place the engine with one side on the upper quarter and the reverse lever in the center, block open the starting valves, and open the cylinder cocks and the throttle. If no steam escapes from either the high or low-pressure cylinder cocks, it has been determined that the valves are not leaking. Now close the throttle and remove blocks from reducing valve, put the reverse lever in about half forward gear and allow all steam previously admitted to the pipes to escape from the open cylinder cocks. Again open the throttle and, as steam will then be admitted behind the high-pressure piston only, a blow through the forward low-pressure cylinder cock would indicate that this packing was blowing. The same operation should be repeated for testing the other side.

Should the engine fail to develop its proper starting power in full gear, what might be wrong? Open the cylinder cocks, with the throttle open and the engine standing, and if steam at no considerable pressure escapes from the low-pressure cylinder, examine the attachments connecting the reverse lever with the starting valves; they operate similarly to the ordinary cylinder cock arrangement and may likewise become defective and thus fail to open the reducing valve.

The compound arrangement for two-cylinder locomotives built by the Cooke Locomotive Works is such that, by means of a small steam

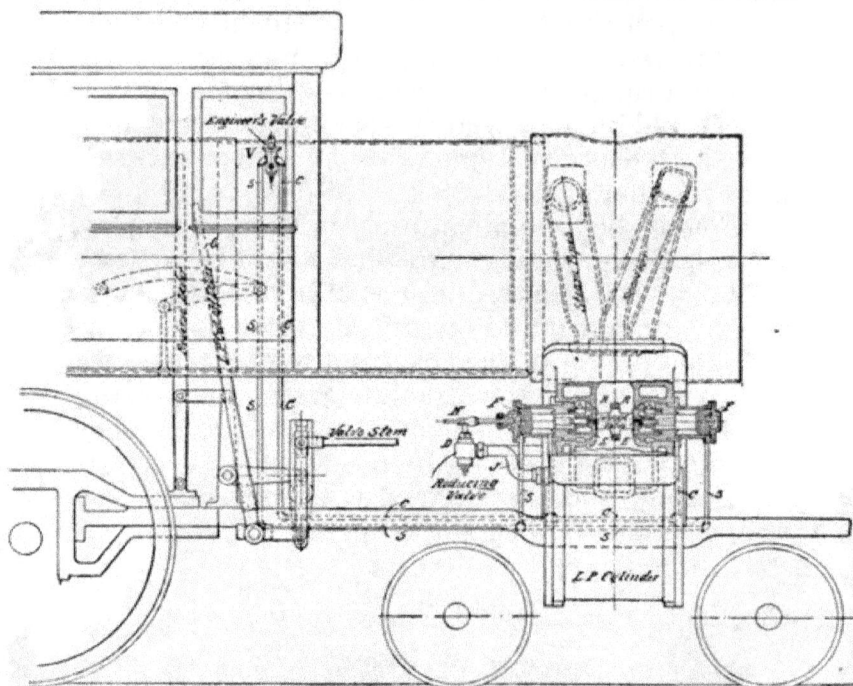

FIG. 174. COOKE COMPOUND.

operating valve in the cab, the engine can be run either simple or compound as the engineer may desire, and is, therefore, a convertible compound. This operating valve admits steam to suitable pistons which open and close the inter-

(134)

cepting valve and operate the reducing valve by a suitable lever connection. The intercepting valve is placed on the right-hand side of the engine within the saddle of the low-pressure cylinder and in the passages *RE*, leading from the receiver to the low-pressure cylinder, as shown in Fig. 174; it also has two connecting ports *AA*, leading into the main exhaust to the stack. By the movement of this valve, steam from the high-pressure cylinder, coming through the receiver, can be thrown out to the atmosphere instead of being allowed to pass to the low-pressure cyl-

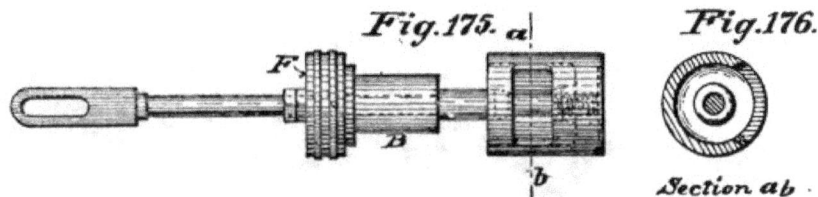

Fig.175. a   *Fig.176.*

Section ab

Cooke Compound
~Back Half of Intercepting Valve~

inder for compound expansion. This movement makes an independent engine of the high-pressure side and at the same time opens a reducing valve *D*, Fig. 174, which supplies, at a reduced pressure, live steam from the boiler to the low-pressure steam chest. The high-pressure cylinder always exhausts into the receiver, and the exhaust from the low-pressure cylinder is direct to the stack with either simple or compound working.

The reader will notice a novelty in the shape

of an additional low-pressure cut-off lever *G*, located in the cab, as shown in Fig. 174. This lever is sometimes applied to Cooke engines of this class and is used to increase or decrease the travel of the slide valve on the low-pressure side independently of the valve on the high-pressure side, in order to equalize the power developed on opposite sides of the engine under variable conditions of steam pressure and different positions of the reverse lever.

Fig. 177

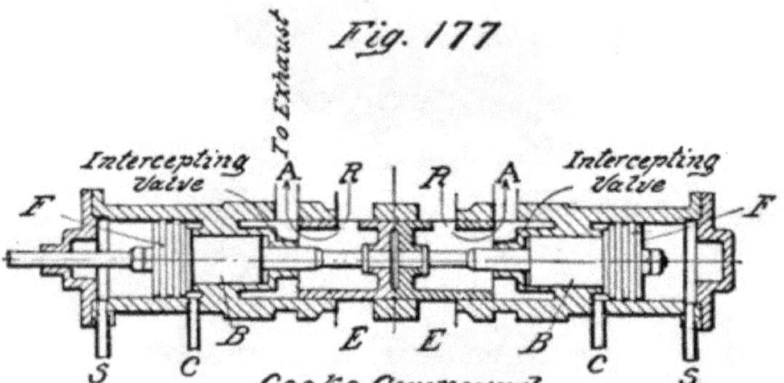

Cooke Compound
When Engine is running simple.

The intercepting valve is composed of two pistons in duplicate, as shown in Figs. 174, 177 and 178, each part being moved by a piston *F* in a separate chamber. Figs. 175 and 176 show one-half of the valve removed from its chamber. It will readily be seen from Fig. 174, that live steam from the engineer's valve *V* in the cab can be supplied through two small pipes *S* and *C* to either the outer or the inner sides, respectively, of these two pistons *FF*. Ports *RR* are passages from the receiver and ports EE lead to the low-pressure steam chest.

To start the engine with single expansion or to run it thus at any time, the engineer pulls the handle of his engineer's valve V in the cab to the back notch. This admits live steam through the small pipe S (Fig. 174) to outside ends of the two pistons FF', thereby moving them and their attached intercepting valve pistons until the two latter come together, the position in which they

Fig. 178

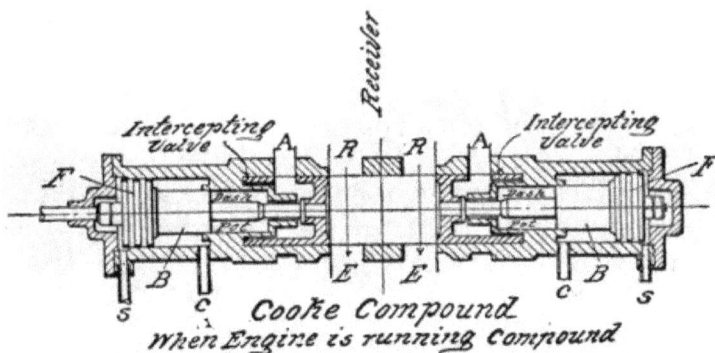

Cooke Compound
When Engine is running Compound

are shown on a larger scale in Fig. 177. In this position the intercepting valve has closed the ports EE to the low-pressure cylinder, and the high-pressure exhaust into the receiver can pass out from ports RR, through the hollow intercepting valves and escape to the atmosphere by exhaust ports AA leading to the main exhaust, as indicated by arrows. The extended and slotted stem of the left-hand intercepting valve piston F (shown in Figs. 174 and 175) in moving to the right has

pulled with it the crank lever K (Fig. 174) and
opened the reducing valve D which, by a connecting pipe tapped into the main steam passage
of the high-pressure side, admits live steam to
the low-pressure steam chest through pipe J.

When the engineer desires to work compound,
he pushes the handle of the engineer's valve V in
the cab to the forward notch, thereby admitting
steam through pipe C (Fig. 174) to the inner
sides of the two intercepting valve pistons FF
and forcing them and their valves outward, as
shown in Fig. 178.   In this position the passage
is opened from the receiver through the intercepting valve ports RR and EE to the low-
pressure steam chest (See Fig. 174).   The outward movement of the pistons FF has shut off the
admission of live steam to the low-pressure cylinder by closing the reducing valve D, and the
steam must now pass from the high-pressure cylinder through the receiver and the low-pressure cylinder to the exhaust or, in other words, the engine
works compound.

Each stem between the operating pistons FF
and the intercepting valve pistons has an enlargement B (Figs. 175, 177 and 178) which fits
loosely into a chamber of the intercepting valve
bushing and thus forms an air dash-pot, thereby
preventing the slamming of the valves when
their positions are changed.

The stem of the piston F which extends back
to operate the reducing valve D is slotted out, as
shown in Fig. 175, so as not to engage the valve
crank K (Fig. 174) until after the intercepting

valve has closed the ports *EE*. Thus the live steam used in the low-pressure cylinder in starting cannot reach the high-pressure side through the receiver and produce a back pressure on the high-pressure piston.

The builders state that in case of accident to either side of the engine, the opposite side may be run for any length of time as a simple engine by disconnecting and blocking the injured side in the same manner as for single-expansion locomotives.

The Dickson Locomotive Works' compound is built under the Dean patents which cover special valves both for automatic and for convertible compounds, but, inasmuch as the practical information is based on the mechanism for the former class only, the detailed description will be confined to the automatic compound.

The starting and intercepting valves are placed on top of the high-pressure steam chest on the right side of the engine. Upon opening the throttle for starting, live steam is admitted to both cylinders, but, after a stroke or two, the intercepting valve automatically opens and the engine works compound thereafter.

The high-pressure exhaust port Q (Fig. 179) is in the balance shield of a Richardson balanced valve P, having its top removed, and thus the exhaust steam from the high-pressure cylinder passes up through it and the intercepting valve G to the receiver and low-pressure cylinder, when the intercepting valve G is open. Beneath the seat R, intercepting valve G, is a port E^1 leading to the chamber E. ·

The receiver, as usual with cross-compounds, is located in the smoke-box, but its shape is out of the ordinary. It is made very large and, between its connections with the high and the low-pressure saddles, branches into two forks, each of which is oval and has metal ribs lengthwise with the pipes. Fig. 180 shows a section through this double portion of the receiver. The object of the designer has been to obtain a very large heating surface, so as to re-evaporate some of the water

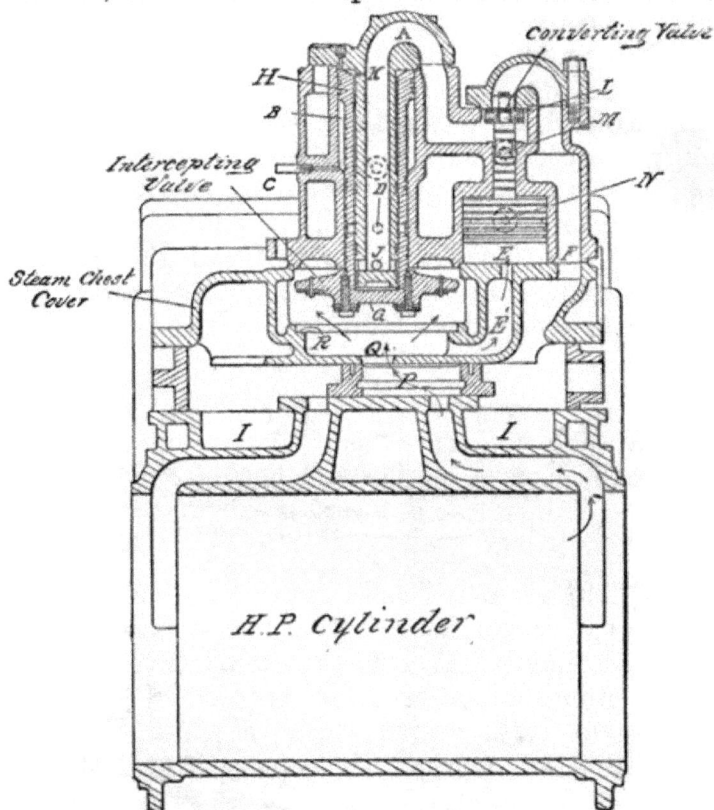

Fig. 179.

Dickson Compound.

condensed in the high-pressure cylinder as it passes with the exhaust steam through the receiver to the low-pressure side. From the reported economy of these engines, it would seem that the object has been largely attained.

Referring to Figs. 179 and 181, it will be seen that the intercepting valve *G* is fastened to the annular stem *H* having an enlarged top above the space *B* which is constantly filled with air pressure or live steam from the pipe *C* when the engine throttle is open, and hence sleeve *H* will be found at the top of its travel, as illustrated, after the engine has started.

The operation is as follows: Open the throttle for starting and live steam enters the high-pressure steam chest from the induction ports *I*

Fig. 180.

—*Dickson Comp*^ª —
— *Cross Section through Receiver*—

(Fig. 179) as usual and besides has a connection to *F* through the top of the steam chest, as shown.   There being no pressure in the receiver (to which chamber *E* is connected through the open intercepting valve *G*), the weight of *N* will have caused the converting valve *L* to drop down, and thus the live steam passes through valve *L* into the tube *AJ*.   Port *K* admits steam to the enlarged top of the annular stem *H*, forcing the intercepting valve *G* down on its seat *R* and bringing the ports *D* in the stem *H* opposite the ports *J* of the central steam tube *A*, thus admitting live steam through them to the

receiver and low-pressure steam chest. With the
intercepting valve G closed (down), the first
high-pressure exhaust, acting in chamber E'
through the port E^1, causes piston N to lift and
close the converting valve L (as shown in Figs.
179 and 181), thereby shutting off the supply of
steam from F to the central tube AJ. What

Fig. 181.

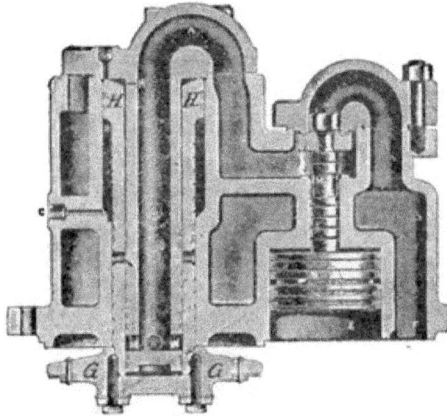

Dickson Compound
Starting Valve in position, working Compound

steam remains in this tube escapes through the
relief port M and allows the intercepting valve G
to move up (open) by live steam pressure from
the pipe C acting in the annular cavity B, as
hereinbefore described, assisted by the high-
pressure exhaust below G. The engine then
works compound, as live steam is shut off from

the low-pressure cylinder and the exhaust from the high-pressure cylinder takes its place.

The engine above described is an automatic compound, that is, starts with live steam in both cylinders but after the first stroke changes automatically to compound.

The inventor of this system, in a design not shown, introduces a reducing valve in the central tube *A* and adds a separate exhaust valve operated by live steam from a three-way cock in the cab, thus making a compound of the convertible class, but at the same time he does not advise convertible construction in compounds.

Accidents to Dickson Compounds.—What should be done in order to run the engine in with the low-pressure side only? Nothing different from a simple engine, but the boiler pressure carried should be reduced about one-half or else the engine throttle opened very slightly.

How could the engine run with the high-pressure side only? There being no means of exhaust except into the receiver, the low-pressure valve would have to be placed so as to uncover the exhaust port or, if that were found to be impossible, the valve entirely removed.

The type of compound locomotive built by the Rhode Island Locomotive Works is sometimes known as the "Batchellor" system, that being the name of the inventor of the device.

In the saddle of the low-pressure cylinder on the left side of the locomotive is located an intercepting and a reducing valve and in the smokebox a separate high-pressure exhaust valve. When the throttle is opened, the engine starts with live steam in both cylinders. With the separate exhaust valve closed the engine automatically changes to compound in the course of a complete revolution; with it open, the engine continues to work as a simple engine as long as desired. This separate exhaust valve is operated at the will of the engineer by means of a three-way cock in the cab; and thus the engine belongs to the class of convertible compounds.

Fig. 182 shows a vertical section lengthwise through the intercepting valve, with the latter in the position when the engine is either starting or being run as a single-expansion locomotive. Fig. 183 shows the same section with the intercepting valve in compound position. R is the receiver port; S is a connection from the main steam pipe; L is a port leading to the low-pressure steam chest, and B is a reducing valve. The intercepting valve is composed of the four pistons, 1, 2, 3 and 4, of which the last works in an oil dash-pot C.

If the engine had stopped after running compound with the valve, as in Fig. 183, and the engine throttle were then opened, live steam from the pipe *S* would force the intercepting valve into simple position as shown in Fig. 182, because piston *2* is larger than piston *1*. In this latter position small port *D* is open and steam from *S* passes through it and the reducing valve *B* to the low-pressure side. Piston *3* has now closed the communication with the receiver *R* in which one

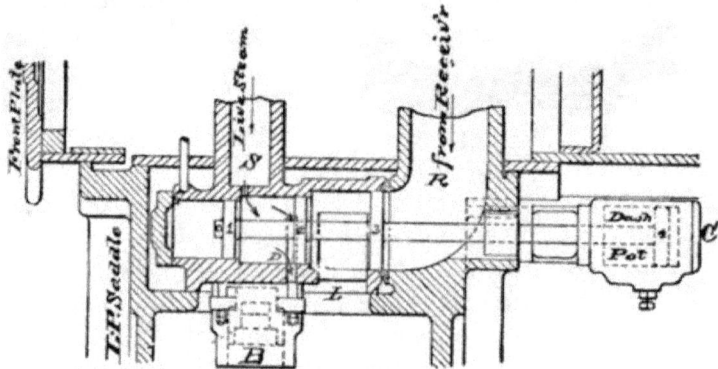

Fig. 182.
Rhode Island Compound
Engine Working Simple.

or two exhausts from the high-pressure cylinder soon produces sufficient pressure to react on this piston *3*, bearing the intercepting valve to the left against the differential pressures on pistons *1* and *2* acting in the opposite direction, and the valve is shifted to compound position, as in Fig. 183, in which position no more live steam can pass through port *D* to the low-pressure side, and the receiver R is connected through port L with the low-pressure steam chest, for which it forms the supply thereafter, as indicated by the arrows.

The port leading from the live steam supply S into the intercepting valve, is larger than it appears from the illustrations, as it extends partly around the circumference of the valve.

The separate exhaust valve shown in Figs. 184 and 185 is placed on the receiver in the smoke-

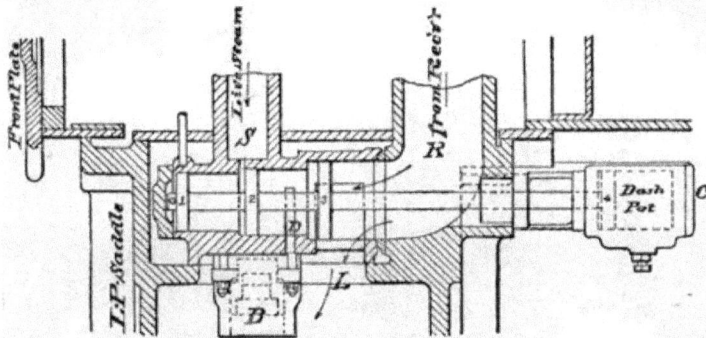

Fig. 183.
Rhode Island Compound
- Engine Working Compound -

box and, when opened by pressure through pipe P leading from a three-way cock located in the cab and under the control of the engineer, connects the receiver with the main exhaust pipe. The opening of this valve will thus permit the high-pressure exhaust to escape and there will be no accumulation of pressure in the receiver. Hence, from the previous explanation, it will be seen that the intercepting valve remains in sim-

ple position (Fig. 182) until such time as the
engineer closes the separate exhaust valve, when
a stroke of the engine automatically changes the
mechanism to compound position, as before described when starting.

The operation of the separate exhaust valve,
shown in Figs. 184 and 185, is very simple. Pressure admitted from the cab through pipe P,
moves the valve V from its closed position (Fig.
185) to the right and vents the receiver pressure
direct to the exhaust pipe, as indicated by the

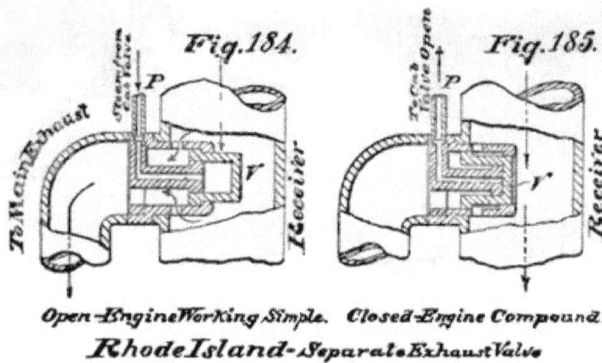

Fig.184.   *Fig.185.*

Open—Engine Working Simple.   Closed—Engine Compound.

Rhode Island—Separate Exhaust Valve

arrows, Fig. 184. Withdrawing the pressure
from the pipe P, allows the receiver pressure to
automatically move and hold closed the valve V,
as in Fig. 185.

Accidents to Rhode Island Compound.—If it became necessary to disconnect either side, how
should the engine be run? Disconnect properly,
observing the same precautions advised for simple engines, then open the separate exhaust

valve so that no pressure can accumulate in the receiver.*

What would you do with a broken intercepting or reducing valve? Open the separate exhaust valve and run with very light throttle, or, preferably, carry a reduced boiler pressure.

Would the working of the engine be affected if the separate exhaust valve V were broken? Probably not; but, if it left an opening between the receiver and the exhaust, the engine would run as a simple locomotive only.

---

*As remarked elsewhere in relation to designs having a separate exhaust valve, if it were the high-pressure side that was disconnected, it would not be necessary to open the separate exhaust valve unless there was some leakage of steam into the receiver.

THE SCIENCE OF RAILWAYS.

IN TWELVE VOLUMES.

BY MARSHALL M. KIRKMAN.

REVISED AND ENLARGED EDITION.

"THE SCIENCE OF RAILWAYS" DESCRIBES THE METHODS
AND PRINCIPLES CONNECTED WITH THE EQUIPMENT,
SHOPS, ORGANIZATION, LOCATION, CAPITALIZATION,
CONSTRUCTION, MAINTENANCE, OPERATION
AND ADMINISTRATION OF RAILROADS.

This great work is everywhere commended for its
thoroughness, vast research and impartial represen-
tation. While it treats of specific things, it does not
reflect the methods of any particular property or country.
A treasury of research and practical experience, it por-
trays truly and vividly the principles and practices of
the great art of transportation in their highest and best
forms. It is popular in every place where railroading
has reached its highest development, and the endorse-
ment it has received from railway men of the highest
attainments is conclusive evidence of its value and trust-
worthiness.

PUBLISHED BY
THE WORLD RAILWAY PUBLISHING COMPANY,
CHICAGO, ILL.
(540)

"ORIGIN AND EVOLUTION OF TRANSPORTATION."

BY

MARSHALL M. KIRKMAN.

[NOTE.—The first chapter in this volume on the "Evolution of Man," has been incorporated since the commendatory notices of the distinguished men and women named below, were written. Originally the volume embraced only the chapters on the Ancients, and the pictures that accompanied the same.]

"For originality of design and thorough treatment of its subject, it is unique among books. Disraeli would have enshrined it among his 'Curiosities of Literature' as a stroke of genius."—Right Reverend WILLIAM E. McLAREN, D. D., D. C. L., Bishop of Chicago.

"It is a work that has great value as a conspectus of the nomadic and other tribes of men, including their contrivances for locomotion. It fascinated me, and I spent the evening poring over its wonderful contents, which are most instructive as well as curious. Every great school, and every college in the world, should possess a book so peculiar and the result of so much research."—Right Reverend A. CLEVELAND COX, D. D., LL.D., late Bishop of Western New York.

"It was a most happy thought that conceived such a work, and in its execution it becomes a most instructive and suggestive contribution to our best literature."—Right Reverend HENRY C. POTTER, D. D., LL D., D. C. L., Bishop of New York.

"It treats well and artistically a comparatively new field of literature." —His Eminence JAMES CARDINAL GIBBONS, D. D., Archbishop of Baltimore.

"The Classical Portfolio is remarkable for the exact historical and geographical information which it imparts, and for the high art which is visible in its engravings. I can not but marvel at the patience, the vast erudition, the literary skill, the esthetic taste of which its pages from first to last give such indisputable evidences. The book instructs and delights. No sooner is it seen than it is appreciated."—Most Reverend JOHN IRELAND, D. D., Archbishop of St. Paul.

The august rulers of the world find the Portfolio quite as fascinating as do their more simple brethren. Love of art, it is thus apparent, knows no degree of rank. His Majesty, King Humbert of Italy, expresses, through his Minister of State, the pleasure the Portfolio has afforded him, and his high appreciation of its great merit. His Majesty, Leopold, King of Belgium, through his Chief of Cabinet, characterizes the Portfolio as a most remarkable and interesting publication and one that elicits his great admiration. His Majesty, the King of Greece, says of the Portfolio that it is at once an artistic, interesting and magnificent production. Many others have been so kind as to express their thanks to the author for the pleasure this rare and beautiful work of art has afforded them. Among the more exalted of these may be mentioned Her Most Gracious Majesty, The Queen of England, and His Imperial Majesty, The Czar of Russia.

"A most interesting and valuable work."—NELSON A. MILES, Major General, Commanding United States Army.

"The work is a splendid one, both from artistic and literary points of view."—WESLEY MERRITT, Major General, U. S. A.

"Mr. Kirkman's researches have taken him into every quarter of the globe, and among every race of man. The illustrations depict every known method of carriage, . . . besides pictures of ancients, medieval and mythological means of carriage. . . . The value of this vast collection is greatly enhanced by the explanations and the dates that are affixed to most of the illustrations, and by the brief historical essays that are prefixed to the several subdivisions."—New York Daily Tribune.

"A superb volume, original in conception and unique in literature and art."—JOSEPH MEDILL, Proprietor Chicago Tribune.

"One of the most exhaustive and valuable books in the world in any given line. It is impossible to say too much in praise of this noble work. . . . Mr. Kirkman is a railroad man of great ability, but he is more, and that is a distinct benefactor to the learning of the world."—The Boston Traveler.

"A more interesting series of illustrations it would be difficult to imagine, or one that could give more clear and positive instruction in the history of humanity. Mr. Kirkman is a distinguished railroad man, and thus he came naturally to the studies which are embodied in this beautiful and comprehensive volume."—New York Sun.

"This volume is especially interesting by reason of its presentation of its subject through a progressive series of excellent pictures following the line of historical development."—Dr. TIMOTHY DWIGHT, President Yale University.

"A book prepared with great care and cost, and possessed of great merit in point of beauty; it is exceedingly interesting and instructive. I am glad to add my testimony to that of others with respect to its excellence."—Dr. FRANCIS L. PATTON, President Princeton College.

"Truly unique, instructive, beautiful and fascinating. I consider it a valuable contribution to literature."—HENRY WADE ROGERS, LL.D., President North-Western University.

"A magnificent volume, . . . most interesting and admirable. It is an illustrated study of the means of travel and transportation in every country in the world that has any; and from the most primitive times down to the present. The profusion of illustrations and their beauty render the work a mine of instructive enjoyment, and I find it interests all who see it."—JENNY CUNNINGHAM CROLY.

"Artistic and valuable, and, despite all traditionary statements to the contrary, something new under the sun."—ELIZABETH BOYNTON HARBERT.

"Exceedingly beautiful, and worthy of all the best words that can be said in its behalf."—MARY LOWE DICKINSON.

PARTICULARS IN REGARD TO THE SALE, DELIVERY, PAYMENT, ETC., OF THE

"SCIENCE OF RAILWAYS."

The work may be paid for in monthly installments. In such cases the books will be delivered as fast as paid for.

In all cases, however, the publishers reserve the right to deliver the books in advance of payment if they so elect.

Payments may be made according to the following plans, as may be agreed upon, viz:

1. In one payment.

2. By remitting draft or postoffice order to the publishers (or such person or persons as they may direct), for the amount of the installments.

3. In the case of those in the service of corporations, orders on the proper officer thereof will be accepted when the publishers are able to arrange with the company to collect the same.

Only those who are duly authorized in writing can collect money on account of the work.

The Publishing Company has had this edition electrotyped. Those who wish all the books delivered at one time may, therefore, expect to receive them without sensible delay. Those who wish them in single volumes from month to month may also confidently anticipate delivery at the time specified.

The work is sold only by subscription.

The World Railway Publishing Company,
79 Dearborn St., Chicago, Ill.